The Enclave Economy

Urban and Industrial Environments

series editor: Robert Gottlieb, Henry R. Luce Professor of Urban and Environmental Policy, Occidental College

For a list of the series, see page 209.

The Enclave Economy

Foreign Investment and Sustainable Development in Mexico's Silicon Valley

Kevin P. Gallagher and Lyuba Zarsky

The MIT Press
Cambridge, Massachusetts
London, England

For information on quantity discounts, email special_sales@mitpress.mit.edu.

Set in Sabon by The MIT Press. Printed and bound in the United States of America.

Library of Congress Cataloging-in-Publication Data

Gallagher, Kevin, 1968–
The enclave economy : foreign investment and sustainable development in Mexico's Silicon Valley / by Kevin P. Gallagher and Lyuba Zarsky.
   p.   cm. — (Urban and industrial environments)
Includes bibliographical references and index.
ISBN 978-0-262-07285-4 (hardcover : alk. paper) — ISBN 978-0-262-57242-2 (pbk : alk. paper)
1. Guadalajara (Mexico)—Economic conditions.   2. High technology industries—Mexico—Guadalajara.   3. Information technology—Mexico—Guadalajara.   4. Investments, Foreign—Mexico—Guadalajara.   5. Sustainable development—Mexico—Guadalajara.   I. Zarsky, Lyuba.   II. Title.

HC138.G8G35   2007
330.972'35—dc22   2006034569

10 9 8 7 6 5 4 3 2 1

Contents

Introduction 1

1 The Promise of FDI for Sustainable Development 13

2 The Emergence of Mexico's Enclave Economy 43

3 Globally Networked, Environmentally Challenged: A Profile of the IT Industry 71

4 Wired for Sustainable Development? IT and Late Industrialization 99

5 Mexico's Bid for a Place in the Global IT Industry 121

6 Silicon Dreams, Mexican Reality 139

7 Importing Environmentalism? 159

8 Beyond the Enclave Economy 177

Notes 193
Bibliography 197
Series List 209
Index 211

Preface

This book went to press on the eve of the tumultuous presidential election in Mexico in July 2006. A pro-globalization champion of a newly emergent middle class squared off against a populist defender of the "losers" of globalization—farmers, workers, the urban poor. For weeks after the razor-thin election, millions of people marched and camped in Mexico City's *zocalo*, convinced that the election had been stolen. The fundamental axis of political conflict in Mexico—and indeed in much of Latin America—involves the issues that we explore in this book: the social, environmental, and economic impacts of "Washington Consensus" policies based on global market-led growth, the role of foreign direct investment in promoting economic development and industrial transformation, and the economic and environmental sustainability of market-driven globalization as a development path.

In the 1990s, Mexico was a "poster child" for globalization. In an abrupt about-face, Mexico threw open its borders to trade and foreign investment, embraced the North American Free Trade Agreement (NAFTA), and ejected government from its role in building up domestic industry. Multinational corporations swarmed into Mexico, creating low-wage jobs in enclaves. Domestic firms, including some in the high-technology sector, went bust at a rapid clip. Along with large numbers of farmers displaced by agricultural imports from the United States, the result was ever-increasing unemployment and migration, a deeper and more apparent gap between globalization losers and winners, and the political mobilization of the "losers" and their allies.

The story was repeated throughout Latin America. The election cycle of 2005–2006 was widely seen as a referendum on "free-market" policies associated with the Washington Consensus. Center-left and left-wing candidates pointed to the Washington Consensus as the reason for the failure of per capita income to rise in 20 years and the persistence of poverty, inequality, low economic growth, and environmental degradation. Their center-right and right-wing opponents countered that the problem lay in not following Washington Consensus policies closely enough. They argued that greater liberalization would generate more foreign investment, which would, in turn, drive economic growth, create employment, and reduce social disparities.

Virtually every presidential contest was very close. In Argentina, Bolivia, Brazil, Chile, Uruguay, and Venezuela, critics prevailed. Only in Mexico, Colombia, and Costa Rica did "neo-liberal" candidates win, and by very slight margins.

There was no closer call than in Mexico. Manuel Andres Lopez Obrador, former Mayor of Mexico City and presidential candidate for the center-left Party of the Democratic Revolution (PRD), criticized Washington Consensus policies and NAFTA in particular as the cause of Mexico's economic perils. Lopez Obrador promised to renegotiate NAFTA's agricultural provisions and steer large amounts of government resources toward the poor. Lopez Obrador's opponent, Felipe Calderon, a former energy secretary under President Vicente Fox (2000–2006) acknowledged that growth and poverty alleviation have been slow to come to Mexico under NAFTA. However, he argued that Lopez Obrador's program would undermine economic stability—a major concern of a newly prosperous middle class—and that the solution lay in increasing foreign investment through even greater liberalization. Despite their differences, the candidates were in broad agreement about the basic "merits" of economic integration in North America and the importance of foreign investment. While he argued for restraints on NAFTA, Lopez Obrador did not—as his critics implied—call for a fundamental change in direction for Mexican economic policy. Indeed, neither candidate had much to say, in public at least, about how to promote the growth of Mexican firms, build an internal market in Mexico, or work in new ways with multinational corporations to boost Mexico's prospects for technological upgrading.

Calderon won the election by fewer than 300,000 votes out of nearly 40 million votes cast. Even if the Mexican Presidential Commission had agreed to a recount, it is unlikely that the margin of victory would have changed much, though perhaps the candidate would have. The political reality is that neither candidate could have gotten a clear mandate to either slow or accelerate Mexico's economic model.

If current trends continue, however, it is likely that they will exacerbate the deindustrialization of Mexico and widen the gap between a foreign enclave economy and the poor and unemployed who stand outside it. The articulation of a compelling and sustainable development strategy—drawing, we hope, in part from this book—may be the key to winning a political majority and putting Mexico's economy on the right track.

We are grateful to the many people and organizations who helped make this book possible. Francisco "Paco" Aguayo Ayala of the Program for Science, Technology, and Development (PROCIENTEC) at El Colegio de Mexico was a wonderful collaborator and friend, helping us formulate our research questions and accompanying us to Guadalajara for two field investigations. PROCIENTEC's Alejandro Nadal provided a forum for outstanding input from Mexican economists at a conference on New Economic Pathways for Sustainable Development in Mexico and remains a great friend and colleague.

Many thanks to our dynamic research assistants, Karen Coppock and Tamara Barber, graduate students at the Fletcher School of Law and Diplomacy, for help in conducting literature reviews in English and Spanish, assembling data, arranging interviews, and preparing the manuscript.

We are greatly indebted to the excellent and generous Mexican scholars of Silicon Valley South. Enrique Dussel Peters directed a comprehensive study of Guadalajara in the late 1990s which generated the edited volume *La Industria Electronica en Mexico: Problematicas, Perspectivas, y Propuestas*. Enrique and other project contributors, including Guillermo Woo Gomez, Juan Jose Palacios Lara, Raquel Partida Rocha, and Maria Isabel Rivera Vargas, shared with us not only their written work but also insights, ideas, and pointers to the right people to interview.

Although we cannot individually name them because they agreed to interviews based on anonymity, we also would like to thank the more than 100 company representatives and government officials we interviewed. Thanks also to the non-governmental organizations who provided insights and information, especially the Center for Reflection and Labor Action, Greenpeace Mexico, Silicon Valley Toxics Network, and the Catholic Agency for Overseas Development.

Our colleagues at the Global Development and Environment Institute at Tufts University provided intellectual and moral support through every phase of the project. Tim Wise of the Globalization and Sustainable Development Program cheered us on with insight, wit, and probing questions. Minona Heaviland and Joshua Berkowitz helped with production design. Colleagues and friends Frank Ackerman, Neva Goodwin, Jonathan Harris, and Julie Nelson provided friendly but sharp advice during institute seminars where we first shared our preliminary findings.

Paco Aguayo, Manuel Pastor, and Eva Paus read the manuscript and provided useful comments and criticisms.

We also received constructive comments during presentations we gave at Harvard University's David Rockefeller Center for Latin American Studies, at the University of California's Center for Latin American Studies (at Berkeley), at Brown University's Watson Institute for International Affairs, and in the Science, Technology, and Public Policy Program at Harvard's John F. Kennedy School of Government.

Two friends and colleagues who deeply inspired our thinking and research on this project passed away during the writing of this volume: Konrad Von Moltke, who pioneered analysis of the relationship between foreign investment and the environment, and Sanjaya Lall, perhaps the foremost economic scholar of the relationship between foreign investment and economic development. Their work lives on in our hearts and on these pages.

We are grateful for financial support from the Rockefeller Foundation, the Charles Stewart Mott Foundation, the Rockefeller Brothers Fund, and Boston University.

We thank Clay Morgan and Robert Gottlieb for their faith in this book. The MIT Press has been rigorous and wonderful to work with on this project from the beginning.

We dedicate this volume to our loving and supportive families. Kevin thanks his wife Kelly and their son Theodore, who was born during the writing of this volume. Lyuba thanks her enduring husband, Peter, and her wonderfully effervescent children, Nadia and Benjie.

Introduction

By the 1990s, Mexico, like many other countries in the Americas, had fully embraced the cluster of policies known as the Washington Consensus: trade and investment liberalization, tight fiscal and monetary policy, the privatization of state-owned enterprises, and in general a reduced role for the state in economic affairs. In Mexico, as in many other countries in the Americas, these policies have had mixed results.

The promise, among others, of following these policies is that foreign direct investment (FDI) by multinational corporations will flow to your country and be a source of dynamic growth. In Mexico, the strategy targeted FDI inflows in manufacturing industries—including information technology (IT)—in order to transform Mexico into the manufacturing base for North America. Beyond boosting income and employment, the hope was that manufacturing FDI would bring knowledge spill-overs (which would build the skills and the technological capacities of local firms, catalyzing broad-based economic growth) and environmental spill-overs (which would mitigate the domestic ecological impacts of industrial transformation).

The strategy appeared to be working, especially after the signing of the North American Free Trade Agreement (NAFTA) in 1994. Now the ninth-largest economy in the world, Mexico was highly successful in attracting FDI, most of it from the United States. Moreover, as was hoped, FDI flowed substantially into the manufacturing sector, including electronics in general and IT specifically, propelling a large increase in manufactured exports, mostly to the United States.

But success has been elusive on other fronts, including creating a substantial number of manufacturing jobs, reducing migration to the United

States, and mitigating the environmental impact of industrialization. Most troublesome for Mexico's long-term development trajectory, the FDI-led strategy has had limited success in stimulating the growth of Mexican firms either as competitors or suppliers to the multinationals. Instead, Mexican industry is being hollowed out and the economy as a whole has been bifurcated into a foreign "enclave economy" and a domestic economy. Within the IT sector, a thriving domestic industry was largely wiped out and replaced by a foreign IT enclave. Moreover, despite the attractions of proximity to the huge US market and the biases toward regional manufacturing in NAFTA, the competitiveness of Mexico as an export platform is being eroded as China's industrialization continues to gain momentum.

This book explores the relationship between foreign direct investment and sustainable industrial development in Mexico through the lens of the IT sector. Our case study focuses on Mexico's "Silicon Valley," an IT cluster in the city of Guadalajara. The central aim of the book is to examine the extent to which increased flows of FDI from the global IT industry helped sustainable industrial development take root in Mexico.

In addition, this book aims to contribute to the larger theoretical and policy debate about the role of FDI in national development strategy. The prevailing view of the Washington Consensus has been that governments should adopt a "hands-off" policy toward markets. In this view, the only rightful roles of government are to enforce contracts, to protect property rights, and to keep the social peace (Kuczynski and Williamson 2003). National capacities for production and innovation—technology, skills, creativity—will then optimally develop exogenously through market forces, including global trade and investment. The best policy for developed countries' governments is to reduce domestic obstacles to the full integration of local firms into the global strategies and supply chains of multinational corporations (MNCs).

The findings in this volume lend credence to a different theoretical framework, one that emphasizes the role of institutions in determining market outcomes. Drawing from traditional political economy, and more recently from evolutionary economics, the institutionalist framework views the development of national capacities for production and innovation as an endogenous process determined by the interaction by governments with market forces. (See, for example, List 1885; Hirschman 1958; Gerschenkron 1962; Amsden 2001; Nelson 2001; Salas 2002; Lall 2005.)

Foreign Direct Investment as Development Strategy

For much of the twentieth century, government was understood to play a catalytic role in economic development. Elaborated in different ways and with varying degrees of success, state-led development was the cornerstone of industrial transformation strategies. The objective was to nurture the emergence of national industries, primarily by mobilizing domestic investment and promoting the growth of a domestic market. A strong case can be made that the most successful economies—Taiwan, Korea, Singapore, and, to a lesser extent, Brazil, Mexico, and India—owed at least part of their achievements to an effective "developmental state" capable of articulating and implementing coherent industrial policies beyond sectional elite interests (Evans 1995; Amsden 2001).

By the 1980s, developmentalist ideas had been replaced by their opposites. A new economic orthodoxy, stemming in part from governments' failures, affirmed the idea that only by excising the promotional hand of the state, attracting foreign investment, and producing for global markets could developing countries alleviate poverty, increase labor productivity, and achieve sustainable economic growth.

Proffered by Organization for Economic Cooperation and Development governments through the World Trade Organization and through multilateral development organizations (most notably the World Bank and the International Monetary Fund), the economic orthodoxy was simple and uniform: developing countries should *liberalize* (that is, open their borders to trade, foreign investment, and global finance, though not to immigrants) and *deregulate* (that is, eschew targeted government promotion of industry while enacting liberal market-friendly policies). More recently, there has also been an emphasis on *good governance* (that is, the creation of institutions that protect contracts and private property).

The engine of development under the globalization orthodoxy is foreign direct investment. Beyond a quick way to boost jobs and income, FDI is considered to be more productive than domestic investment because it generates a variety of positive externalities. "The overall benefits of FDI for developing country economies are well-documented," claims the Organization for Economic Cooperation and Development (OECD 2002). Based on broad consultations with industry, government, and non-government organizations, the report concludes: "Given the appropriate host-country policies and a basic level of development, a preponderance of studies shows that FDI triggers technology spill-overs, assists human capital formation, contributes to international trade integration, helps create a more competitive business environment and enhances enterprise development. All of these contribute to economic growth, which is the most potent tool for alleviating poverty in developing countries." That optimistic assessment, however, is not borne out empirically. The "preponderance" of statistical studies in fact finds no evidence that FDI brings positive spill-overs to developing-country firms in the same industry, and some evidence that the spill-overs are instead negative—domestic firms that compete with MNCs either contract or go out of business. There is some evidence that FDI generates spill-overs to local supplier firms. Case studies, on the other hand, find ample evidence of positive FDI spill-overs—but only when governments play an active role in capturing them.

The FDI "success stories" tend to be largely in East Asia: Taiwan, Singapore, China, and more recently India. Although Korea eschewed FDI, it obtained benefits from multinational corporations through direct licensing agreements. What these countries have in common is a pro-active government that effectively nurtured the capacities of domestic firms to learn and innovate. They did so through a variety of policy and institutional interventions, including targeted industry policies and investment in education, in research and development, and in science and technology training. Latin American governments, in contrast, largely adopted a passive approach, seeking benefits through increases in the quantity rather than the quality of FDI inflows.

Sustainable Industrial Development

A second theoretical contribution of this book with policy implications is to re-conceptualize the goal of foreign direct investment as promoting "sustainable development." Despite a nod to environmental and social parameters, development theory and policy tend to be dominated by a focus on economic objectives, often defined narrowly as increases in gross national product or in GNP per capita. Moreover, rather than development per se, policies directed at developing countries often emphasize poverty alleviation measured solely in terms of income and employment. The purview of this book is the broader concept of sustainable industrial development, which we define as simultaneous evolution along three axes:

Economic an increase in the endogenous capacities of domestic firms and workers to learn, innovate and produce for domestic and/or global markets

Social the creation of jobs, especially for the poor and for the middle class

Environmental mitigation of the environmental and health impacts of industrial growth.

Endogenous capacities for production and innovation held within domestic firms form the bedrock of sustainable industrial development. In most medium to large economies, local firms account for the bulk of industrial activity. An influx of FDI can quickly boost export earnings and local employment. However, if MNCs are disarticulated from the local economy—that is, if financial, technological, knowledge, and human capital transfers are largely kept within foreign companies—local firms do not learn to undertake more advanced technological functions that enable them to move up the value chain. The benefits of FDI are captured primarily within the foreign enclave rather than diffused through the economy.

Moreover, without local diffusion of knowledge and technology, countries will not be able to maintain its competitiveness in the face of mobile FDI. As Lall argues (2005, p. 3), "great domestic development capacity [is necessary] to sustain export competitiveness." In manufacturing industries, the local supply and niche market capacities of local firms are central in the global investment strategies of MNCs. Yet developing countries are often lacking what Maurizio Carbone, Director General of the United Nations International Development Organization, calls "the missing middle": dynamic, highly productive small and medium-size companies, which are "important for a resilient and flexible economic structure" (*Courier* 2003).

Paus (2005) has demonstrated that stable property rights, political stability, and social peace are indeed "location specific assets" that attract FDI to developing countries, but much more is also needed. Cost advantages, including proximity to major markets, and appropriate infrastructure are also important. Moreover, location-specific assets must be aligned with MNCs' strategic interests, which are influenced by global commodity chain dynamics, trade rules, risk assessment, and other factors. Finally, success in attracting high-tech FDI does not guarantee the capture of knowledge spill-overs. To do so, domestic firms must overcome information and coordination externalities and must be cost competitive as suppliers of inputs to MNCs. Of course, MNCs must be interested in purchasing local inputs in the first place (Paus 2005).

The social aspect of sustainable industrial development is often described in terms of promoting economic and political equity. While wholeheartedly supporting these goals, we focus in this book more narrowly on the role of government and FDI in nurturing industries that maximally generate and sustain jobs. Likewise, FDI has a wide array of environmental impacts (Zarsky 2002). In this book, we focus on the role of FDI in generating spill-overs—including cleaner technology and better management practices—that reduce the environmental costs of industrialization.

Spotlight on Mexico's IT Industry

For several reasons, Mexico's experience with the global IT industry offers a fertile laboratory to study the impacts of foreign direct investment on sustainable industrial development in developing countries. First, not all developing countries are successful in attracting FDI in general and in the IT sector in particular. Indeed, ten countries receive more than 90 percent of the FDI that flows to developing countries, and Mexico is one of the top three. Second, successive Mexican governments have articulated not only economic but also social and environmental goals for the FDI-led strategy. Third, the Mexican government identified the IT sector as a priority industry and targeted it with efforts to attract FDI. Fourth, Mexico embraced the Washington Consensus with speed and vigor. In an about-face from its 40-year experiment with state-led industrialization, the Mexican government, in rapid succession, eliminated support for domestic firms and adopted non-preferential "horizontal" industry policies, opened its borders to trade and foreign investment, and, despite earlier misgivings, signed NAFTA. In the electronics sector, a highly successful program to promote domestic manufacturing firms and domestic markets was dismantled. Finally and most important, the strategy worked—but only in the short term. "Global flagships" of the IT industry, including Hewlett-Packard and IBM, flocked to Guadalajara and established manufacturing operations in the 1990s. Initial hopes that local firms would evolve into contract manufactures and suppliers to the MNCs, however, were short-lived. In the mid 1990s, however, global flagships began to outsource manufacturing operations. Rather than contract with local Mexican firms, the global flagships invited large, mostly US contract manufacturing firms (CMs) such as Flextronics and Solectron to co-locate in Guadalajara. The CMs, in turn, built their competitive advantage on managing a third tier of global suppliers, many in East Asia, on a razor-thin profit margin. Less than 5 percent of inputs were sourced locally. Far from generating broad-based growth, Mexico's "Silicon Valley" had been transformed by 2000 into a foreign enclave, and only a few of the original 50 Mexican IT firms were still in business.

The heavy reliance on foreign contract manufacturers and imported inputs meant that the FDI-dominated IT industry generated few backward linkages—an important channel for knowledge and environmental spillovers to Mexican firms. Moreover, since most of what was produced was exported, there were few forward linkages through diffusion to domestic markets. IT products in particular help to increase efficiency in other sectors of the economy. A third channel for spill-overs is through human capital: workers or managers trained on the job may start their own companies or take their skills to other companies, thus increasing the productivity of local firms. In Guadalajara, however, IT operations employed a largely temporary, semi-skilled workforce with little opportunity for skill enhancement.

The lack of backward and forward linkages, as well as the lack of skill growth, meant that FDI in the IT sector generated few knowledge or environmental spill-overs in Guadalajara. In addition, the Mexican government did not develop an adequate environmental regulatory framework for the IT sector, which was considered a low priority compared to other industries. As a result, there was little incentive, apart from pressure by NGOs or by corporate headquarters, to improve environmental management.

Still, the global flagships and the contract manufacturers were a source of employment and income—at least until 2000, when a crisis of overcapacity hit the industry. When the high-tech stock bubble burst, the global flagships scrambled to find cheaper production sites. In 2001, China joined the WTO. With its large domestic market, low wages, and significant IT manufacturing capacities (built up over 20 years by state-led development policies), China became the production platform du jour. Manufacturing operations in Guadalajara were severely cut back or relocated, generating massive layoffs. Left in Guadalajara was a handful of foreign contract manufacturers struggling to maintain orders for low-end assembly operations—and a chorus of local analysts bemoaning a missed opportunity.

The Maquila Mindset

Why, despite early optimism, did the promise of Mexico's "Silicon Valley" go unfulfilled? We demonstrate that Mexico's FDI-led plan for IT development stemmed from the interaction of two factors: (1) the narrow vision and passive industrial policy framework of the Mexican government's pro-FDI development strategy and (2) the dynamics of global restructuring and competition in the IT industry.

The root cause is the Mexican government's overreliance on FDI-led development. During the ISI period, the government failed to deploy reciprocal control mechanisms to ensure that domestic firms could export close to the global technological frontier, thus leaving the domestic firms relatively less efficient when liberalization came. During the liberalization period, Mexico went too far with the belief that it would be sufficient to "let markets do it," Mexico did not implement strategic policies aimed at building the capacities of local firms and workers to learn, innovate, and absorb spill-overs. In addition, Mexico did not—with one exception—build partnerships with MNCs to encourage knowledge transfer. Indeed, the exception proves the rule. In the period under study, a government partnership with IBM generated the only start-up firms.

The experience of other late-industrializing countries, especially in East Asia, is that the state must proactively promote local learning, knowledge, and innovation. With such policies in place, FDI spill-overs can be garnered. Without them and the growth of local knowledge assets they engender, MNCs will transfer only low-skilled, low wage and ultimately footloose operations.

Rather than a proactive industry policy to develop domestic firms and markets, Mexico adopted a "maquila mindset"[1] that oriented industrial development solely around attracting MNCs to produce for export. We see the maquila mindset as an outgrowth of the maquiladora industrialization program of the 1960s, when Mexico attracted MNCs that didn't have to pay import duties. The result was islands of growth with no links to the economy. Indeed, rather than a "level playing field" for foreign and domestic firms, Mexican policy went so far as to (inadvertently) *favor* foreign over domestic firms. There were tax breaks for foreign firms and incentives to import intermediate inputs. Credit to domestic firms dried up, in part because of the constriction of government-provided development financing, while foreign firms had access to global capital markets and their own internal sources of financing. In addition, to attract foreign capital, Mexico kept interest rates high, choking off domestic investment and putting upward pressure on the peso. Imports made cheaper by an overvalued peso further biased MNC procurement away from domestic suppliers.

For their part, MNCs were biased toward foreign firms in outsourcing manufacturing functions and procuring inputs, while the foreign contract manufacturers were biased toward foreign inputs. A commitment to build the capacities of local firms was not—and is not—seen as a part of the mantle of "corporate social responsibility" embraced by many MNCs in the global IT industry.

The MNCs diversified from Guadalajara abruptly when two global conditions changed. The first was the crisis of overcapacity that hit in 2000 with the collapse of the high-tech stock market bubble. The second was the accession of China to the WTO. China was already attractive as a manufacturing platform because of its low wages, huge domestic market, and substantial productive capacities. Its accession to the WTO reduced tariffs and other obstacles to FDI. Every MNC that relocated from Guadalajara headed for China.

Beyond the Enclave Economy

What can Mexico and other developing countries learn from the Guadalajara experience?

The most important lesson is that foreign direct investment is not a miracle drug. Expecting FDI to automatically stimulate economic growth and transform industry—and designing policies accordingly—is more likely to generate enclaves than spill-overs. The Guadalajara experience confirms the findings of other studies: supportive public policies are needed to nurture domestic industries and capture the benefits from FDI. FDI is more likely to support industrial upgrading and promote endogenous productive capacities within a framework of clearly articulated development objectives and supportive public policies.

A second lesson is that relying on low wages to attract MNCs is risky: still lower-wage platforms are just around the corner. In the early 1990s average manufacturing wages in Mexico were low compared to the United States and Canada, but by 2001 they were 4 times the level of wages in China. Lacking knowledge-based assets and the pull of domestic market access, export-oriented manufacturing platforms in developing countries are vulnerable to footloose MNCs.

A third lesson is that the benefits of FDI in the IT sector may be limited for developing countries. The industry is highly concentrated at the top, with a few firms accounting for the bulk of sales—and profits. Though touted as knowledge-intensive and high-tech, MNCs tend—without proactive policies—to transfer low-skilled, low-value manufacturing and assembly operations with low profit margins and few spill-overs. Even with proactive policies, barriers to entry in the highly concentrated global IT industry may be too high for developing countries to overcome. The industry is also highly concentrated geographically, with a large—and growing—portion of production sited in Asian countries.

Methodology and Outline

This book is based largely on field investigations conducted in Guadalajara from October 2003 to June 2005, and follow-up telephone and email interviews conducted throughout 2006. The central question was: Did

FDI foster sustainable industrial development in Mexico, particularly in the prized IT sector? We asked three more specific questions:

• Did Mexico attract and sustain FDI in the IT sector?

• When high-tech FDI came, did it generate productivity spill-overs?

• Did FDI in the IT sector transfer cleaner technologies and environmental standards and spur their suppliers to adopt the practices of such technologies?

We drew on both quantitative and qualitative techniques. We conducted more than 100 interviews with US and Mexican IT companies and with industry associations, government officials, academics, journalists, workers, and non-governmental organizations.

We interviewed corporate officers of Mexican affiliates of Hewlett-Packard, Motorola, Lucent, and Intel; and plant managers and/or Environmental, Health and Safety Officers of Flextronics, Jabil Circuit, and SCI-Sanmina. We also interviewed representatives of the Mexican IT firm Electronica Pantera and of the Jalisco-based Electronics Industry Supply Chain Association. We also spoke with representatives of non-governmental organizations that monitor the IT industry, including Greenpeace and the Catholic Agency for Overseas Development.

We interviewed Mexican government officials at the Environment Protection Agency (PROFEPA), the National Registry of Foreign Direct Investment, and the Ministry of Economy. We also spoke with the state of Jalisco's Secretary of Economic Promotion (SEPROE). Finally, we spoke with a number of Mexican academics who have studied the IT sector in detail.[2]

From many of these sources we were also able to obtain a great deal of quantitative data, some of which is very difficult to obtain otherwise. Having data on broader trends allowed us to be more rigorous in that we could juxtapose personal stories with broader trends. As the book shows, more often than not, both methods tell the same story.

The first two chapters explore the broad theoretical and historical landscape. Chapter 1 explores the gap between the promise of FDI for industrial transformation—the delivery of economic and environmental spill-overs—and the empirical evidence in developing countries. Chapter 2 describes the emergence of Mexico's FDI-led liberalization strategy and analyzes its macroeconomic impacts on growth, manufactured exports, domestic investment, environmental performance, and job creation.

The next two chapters examine the structure, dynamics and economic geography of the global IT industry. Chapter 3 charts the emergence of the "global production network" mode of organization of the IT industry, a mode characterized by the outsourcing of manufacturing functions by global flagship firms to large, highly concentrated contract manufacturers and by numerous suppliers operating on razor-thin margins. Chapter 3 also examines the occupational health and environmental challenges of the IT industry, including the rising import standards of the European Union. Chapter 4 considers the promise and perils to developing countries of trying to enter or to upgrade to a higher level in the global IT industry and describes the experience of successful East Asian producers—Taiwan, Korea, Malaysia, India, and, especially, China.

The next four chapters present a case study of the IT sector in Guadalajara. Chapter 5 outlines the success of early Mexican industry policy in developing an electronics manufacturing capacity and the later initially successful but ultimately failed bid to enter the global IT industry in the context of regional economic integration. Chapter 6 examines why FDI generated so few economic spill-overs, and chapter 7 asks why it generated so few environmental spill-overs. Chapter 8 analyzes the factors that account for failure and the lessons it holds for developing countries.

1

The Promise of FDI for Sustainable Development

Foreign direct investment is at the center of the mainstream global development paradigm. The standard prescription is to "let markets do it" by removing institutional and policy obstacles to inflows of FDI and to operations of multinational corporations. In addition to stimulating economic development and reducing poverty, FDI is increasingly seen as a route to improving the environmental management of industry in developing countries.

In the short term, FDI potentially brings two types of economic benefits. First, it boosts national income and employment Indeed, the promise of jobs is what buys public support for the sometimes exorbitant subsidies used by both developed and developing countries to attract FDI.[1] Second, FDI is a source of external capital, one generally seen by developing countries' governments as less volatile than debt, thus providing ballast for both the current account and the exchange rate.[2]

In the long term, the hope is that FDI will nurture sustainable growth in host countries by transferring technology and upgrading skills. The sustainable growth effect stems from the deployment by multinational corporations of a "bundle of assets" scarce in developing countries, including technology, management skills and systems, global marketing channels, research and development capacities, and brand names (Agosin and Mayer 2000). In theory, these assets make FDI *more* productive than domestic investment.

Although the notion that FDI delivers an "added bonus" to economic development is theoretically coherent, the evidence for it is thin. For more than 20 years, researchers searching for positive spill-overs from FDI to domestic firms in developing countries have found instead that the link is as likely to be negative or neutral.

This chapter considers three kinds of positive externalities by which FDI may potentially deliver an "added bonus" to economic development:

- knowledge spill-overs (transfers of technology and know-how from MNCs that increase the productivity of domestic firms)
- "crowding in" of domestic investment (in which a dollar of FDI adds more than a dollar to gross fixed capital formation by stimulating domestic investment)
- sustainable-development spill-overs (transfers by MNCs of cleaner technologies and "best-practice" environmental management systems).

Global FDI Trends

In the 1990s, FDI soared to unprecedented levels. From an average of $58 billion a year between 1970 and 1990, FDI inflows swelled to an average of $311 billion per year between 1992 and 1997. FDI was $691 billion in 1998, $1.1 trillion in 1999, and $1.4 trillion (4 percent of global GDP) in 2000. Since 2001, FDI inflows have fallen steadily. Nonetheless, the long-term trend is up, with inflows in 2003 of $560 billion (figure 1.1).

FDI is one of three ways foreign capital can flow into developing countries. The other two are bank loans and portfolio investment. FDI is undertaken by multinational corporations, primarily those headquartered in developed countries. MNCs invest overseas through mergers and acquisitions or by creating new companies, usually by establishing a local affiliate (a "greenfield").[3]

To make FDI profitable, a multinational corporation must have some distinctive productive asset not possessed by domestic firms, such as technology, global marketing capacities, access to capital, or management skills (Blomstrom and Kokko 1996). The firm is thus able to earn a "rent." Host countries, at least in theory, receive beneficial "spill-overs."

Developing countries typically lack a wide range of productive assets, suggesting that they would be attractive to rent-seeking MNCs. However, social, political and economic conditions in developing countries subject MNCs—as well as domestic investors—to a high degree of risk. Moreover, private investment, both domestic and foreign, requires a substantial sup-

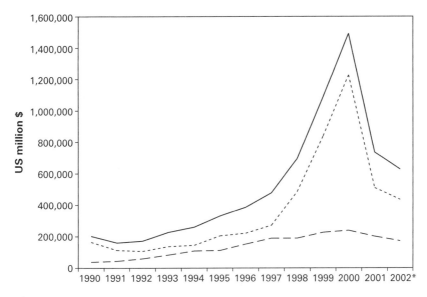

Figure 1.1
FDI inflows in the world economy, 1990–2002. —: world. - -: developed countries. – –: developing countries. Source: UNCTAD 2004.

porting foundation of public infrastructure. Beyond transportation lines and literate workers, requisite infrastructure includes public values and institutions that honor contracts and protect private property.

The combination of risk and lack of infrastructure in developing countries affects the global FDI picture in two ways. First, the majority of the world's FDI—about 70 percent on average in the 1990s—flows from one developed country to another. Second, among developing countries, FDI inflows are highly concentrated: only ten countries get 75–80 percent of the total. The top recipients are China, Mexico, Brazil, and (before the financial crash) Argentina. Indeed, in 2003, China was the second-largest recipient of FDI inflows in the world. The entire continent of Africa received less than 3 percent (UNCTAD 2004, table 2).

Though minuscule in the global picture, FDI inflows and MNC operations in poor or small developing countries may comprise a significant part of the local economy. In Angola, for example, FDI inflows accounted for more than 40 percent of gross fixed capital formation between 1996 and 1998 (UNCTAD 2000). In Costa Rica, a $600 million investment by US chipmaker Intel Corporation was described as "putting a whale in a swimming pool" (Larrain 2000). Intel single-handedly accounted for about 8 percent of GDP in 2000 and more than 25 percent of the country's exports (Luxner 2000; UNCTAD 2002).

For much of the 1990s, manufacturing was the leading sector for FDI in developing countries, accounting for 53 percent of total FDI inflows on average between 1989 and 1991. However, the fastest-growing sector is services, including electricity, gas and water, construction, finance, health and social services, and transport, storage, and communications. By 2001–2002, the service sector accounted for half of FDI inflows to developing countries, up from an annual average of 35 percent in 1989–1991. The proportion of FDI inflows into the primary sector changed little, accounting for about 10 percent of the total.

Determinants of FDI

What motivates MNCs to invest in developing countries? What determines which countries they invest in? Assuming that, in general, MNCs earn rents by investing in developing countries, their investment in a particular country is generally understood to be motivated by a search for one or more of the following[4]:

Resources MNCs are attracted by the opportunity to explore or exploit non-mobile resources, including minerals, petroleum and other raw materials, as well as skilled labor. In addition, countries may have an attractive locale-specific infrastructure asset, such as a port, or technological assets, such as an innovative industry cluster.

Production costs MNCs look for investment opportunities in host countries in which they can lower their costs of production, including wages adjusted for labor productivity. The point is not simply to find the lowest wages but the lowest-cost labor per unit output. Other production costs include transport, intermediate inputs, waste management, and the costs entailed with local regulation, including taxes and standards.

Markets MNCs invest in particular countries to gain access to domestic or regional markets. Access may be problematic via trade because of the presence of tariff or non-tariff barriers to imports, as well as lack of market information, such as consumer preferences.

Before the globalization wave of the 1990s, FDI was understood to be driven primarily by the search for market access, and secondarily, especially in developing countries, the search for resources. With the liberalization of trade and investment, according to UNCTAD (1996), "cost differences between locations, the quality of infrastructure, the ease of doing business and the availability of skills have become more important" than the pull of the domestic market. If true, than countries with small domestic markets could compete favorably with large market countries in attracting FDI inflows seeking efficiency gains.

Some analysts, however, are skeptical that FDI is propelled by a gap in the efficiency between foreign and domestic firms. Krugman (1998) argues that, generally, domestic investors are more efficient than foreign investors in running domestic firms—otherwise, foreign investors would already have purchased them. However, in financial crises, such as the crisis that swept East Asia in the late 1990s, domestic firms may be cash-constrained and be available for purchase at "fire-sale" prices. Krugman concludes that a superior cash position, rather than cost-reducing technology or management, drives FDI (ibid.).

Another team of analysts argues that FDI is driven by the information advantage of foreign investors, who are able to gain—and leverage—inside information about the productivity of firms under their control. With their superior information, foreign firms can inflate the price of equities sold in domestic stock markets. The expectation of future stock market opportunities then leads to overinvestment and inefficiency (Razin 1999).

Moreover, a recent study found that there is "limited evidence" that market-seeking is no longer the primary motivation for FDI. Based on survey data of 28 developing countries, the study found that, during the 1990s, FDI stocks remained highly correlated with "traditional" market-pull factors such as population size and GDP per capita of host countries, as well as administrative bottlenecks (Nunnenkamp and Spatz 2002)

Further support for market-seeking as the motivation for FDI is the fact that the fastest-growing sector for FDI inflows in developing countries is services, including transport, telecommunications, energy, and finance. With some exceptions, such as data processing and software programming, services are aimed at domestic markets.

In addition to domestic markets, FDI may be motivated by a search for access to neighboring or regional markets. For example, Ireland became a platform for MNCs looking for access to European Union markets, while proximity and access to the United States and Canada has augmented FDI inflows into Mexico.[5] Understanding FDI as motivated primarily by market-seeking helps to explain why more than half of the total flowing to developing countries goes to China, Brazil, and Mexico.

Some analysts have suggested two other potential drivers of FDI in developing countries: (1) lower environmental standards (the "pollution haven" hypothesis) and (2) liberal investment agreements. To date, however, the empirical evidence suggests that neither plays much of a role in determining the location decisions of MNCs. In most industries, pollution abatement costs are too low to register (Jaffe 1995). While the environmental *performance* of MNCs may be lower in low-standard than in high-standard countries, there is little evidence that low standards drew them there in the first place (Zarsky 2002).

Eager to attract FDI, many developing countries have signed bilateral investment agreements (BITs) with the United States and other capital exporting nations. These BITs significantly increase protection for private foreign investors, even at the expense of public goods (Zarsky 2005). However, there is little evidence that signing a BIT increases FDI inflows. In a study of FDI from 20 OECD countries to 31 developing countries between 1980 and 2000, the World Bank (2003) found that BITs do not stimulate additional FDI. Rather, the key drivers of FDI are market size and macroeconomic stability.

While the aggregate data point toward market-seeking as the primary determinant of FDI inflows, MNCs are subject to sector-specific and industry-specific pressures. MNCs in the minerals sector, obviously, are pulled by the potential or proven existence of exploitable resources, while MNCs in the manufacturing sector may be pulled by a combination of low wages and access to domestic and/or regional markets. Indeed, China offers advantages on all three counts: mineral and fossil fuel resources, low production costs due to an abundance of low-wage labor, and a huge internal market.

Moreover, MNCs develop firm-specific strategies to gain advantages over competitors. Global strategies include targeting particular markets and selecting host locations, usually out of a plethora of options. For example, Intel carefully considered six other countries in Latin America before choosing Costa Rica as a microprocessor production base for its North American and Pacific Basin markets.[6] Costa Rica was chosen because of its location-specific assets, including tax exemptions, the high educational level of the work force, political stability, and a corruption-free environment (Larrain 2000).

MNCs also make strategic decisions about which functions to establish in which locations, such as whether to develop an R&D capacity in a developing country or to keep it in headquarters; and about how to manage their global supply chains, including which (if any) inputs to source locally versus globally from overseas suppliers.

The motivation and global strategy that draws a multinational corporation to invest in a particular site is likely to exert strong influence on the benefits—or lack of them—reaped by host countries. If FDI is motivated by a search for low wages, transfers of state-of-the-art technology and research and development functions are unlikely. Likewise, if a multinational corporation has a strong bias toward global suppliers, it is unlikely to generate spill-overs through backward linkages with domestic firms.

Knowledge Spill-Overs

Spill-overs are leakages of knowledge, intended or unintended, from one firm to another. In the case of FDI, they are knowledge leakages from MNCs to local firms. The concept of knowledge spill-overs encompasses both technology and "tacit" knowledge, that is, know-how as applied to any aspect of production, including management.

Spill-overs are externalities in the sense that the recipient obtains knowledge without paying the full cost of inventing or creating it. Central to the concept of spill-overs is that the recipient "captures" the leaked knowledge and uses it for commercial advantage by increasing productivity and/or competitiveness.

Knowledge spill-overs are increasingly understood by economists to be the principal way that FDI stimulates economic growth in developing countries. Through MNC spill-overs, developing-country firms can leapfrog to higher productivity and become globally competitive.

Active spill-overs involve the direct transfer of knowledge by MNCs; for example, the transfer of blueprints for a production technology or a management system, including through licensing arrangements and joint ventures. *Passive* spill-overs involve embodied knowledge—the employment of specialized and advanced intermediate products. The primary transmission belts for passive spill-overs are FDI and international trade.

For "free-riding" firms in developing countries, the prospect of gaining commercially useful knowledge without the large investments in R&D required to create it themselves makes FDI highly attractive. MNCs, on the other hand, generally seek to prevent or minimize leakages in order to protect the rents generated by their proprietary knowledge. However, some forms of firm-specific knowledge have ceased generating rents and MNCs may be quite willing to share it. Moreover, sharing knowledge with employees in foreign subsidiaries and downstream suppliers is an operational requirement for MNCs. Thus knowledge may leak out despite incentives to keep it within the company. Host-country spill-overs from FDI can potentially be captured by MNC subsidiaries, by other firms in the same industry as the MNC (horizontal spill-overs), by downstream suppliers to the MNC (vertical spill-overs), or by firms in upstream and other industries. Except for MNC subsidiaries, whose access to knowledge is directly determined by their corporate parents, spill-overs may occur in five ways:

Human capital   MNCs hire and train both skilled and unskilled workers. If they leave the MNC, skilled workers can apply their knowledge either in starting their own firms or in working for domestic firms in the same industry. If their knowledge improves productivity in the new environment, a spill-over has occurred. Likewise, unskilled or lower-skilled workers can apply their knowledge to other firms in the same industry, potentially with higher technology content. To retain their workforce and prevent human-capital spill-overs, MNCs have an incentive to pay higher wages than local firms.

Demonstration effects   Domestic firms may adopt and produce technologies introduced by MNCs through imitation or reverse engineering. They may also adopt higher, productivity-enhancing standards of MNCs in relation to inputs, quality control, environmental management and labor. Governments may codify these standards in new regulation, spawning further gains in efficiency.

Competition effects   Except in sectors where there are no indigenous firms or where they are offered a monopoly status, MNCs compete with local firms for domestic and export markets. The presence of MNCs may exert price pressures on domestic firms to adopt new technology or to utilize existing technology more efficiently.

Backward linkages   Domestic suppliers of intermediate inputs to MNCs may capture spill-overs through technical training to meet specifications, as well as requirements and training to meet global standards (Hirschman 1958). If MNCs purchase a substantial volume of inputs locally, and/or if they help their local suppliers find additional export markets, they may also allow local suppliers to capture economies of scale, thus increasing productivity and potentially fostering the "crowding in" of domestic investment.

Forward linkages   MNC-produced goods and services may enter into and increase the productivity of production processes of firms in upstream and other industries. Information and communication technologies, for example, are important inputs to virtually all industries.

While the hope is that knowledge spill-overs will generate a "virtuous circle" of higher productivity and economic growth, it is possible that FDI will instead have negative externalities. Negative *horizontal* spill-overs occur when domestic firms in the same industry as MNCs become less productive and shrink in size or go out of business. Negative *vertical* spill-overs occur when domestic downstream firms are displaced from the market by MNC biases toward foreign suppliers (Chudnovsky 2004).

Evidence for Knowledge Spill-Overs

Since the 1970s a large literature has emerged to determine empirically whether and how FDI generates knowledge spill-overs, both in developed and developing countries. In broad terms, the studies are of two types: statistical studies and case studies.

Focused exclusively on the manufacturing sector, the statistical studies seek to find statistically significant links between concentrations or inflows of FDI and the performance of domestic firms in terms of labor or total factor productivity or exports. [7] A higher propensity of domestic firms to export is seen as an increase in their efficiency and global competitiveness. Most statistical studies focus on *horizontal spill-overs*—impacts on domestic firms in the same industry as the MNCs. More recently, researchers have begun to look for *vertical spill-overs*—impacts on downstream or upstream domestic firms.

Case studies, especially in developing countries, tend to examine how FDI has helped or hindered the emergence and/or evolution of globally competitive domestic firms. While they incorporate data on productivity and exports, case studies consider more broadly how the interaction of FDI with local institutions, policies and firm capacities has worked—or not worked—to nurture the growth not only of pre-existing domestic firms but of whole new industry sectors. Some case studies focus exclusively on intra-industry impacts, that is, horizontal spill-overs, others on how FDI has affected local supplier firms, and still others on both horizontal and vertical spill-overs.

Horizontal Spill-Overs

Using cross-sectional, industry-level data in a single year, a number of early studies found that industries with a higher concentration of FDI were more productive (Caves 1974; Blomstrom 1983; Blomstrom and Wolff 1994). Positive spill-overs were found in both developed countries (e.g. Australia and Canada) and developing countries (e.g. Mexico).

Later studies, using both industry and firm-level data, found much more ambiguous and modest results (Kokko 1996; Aitken and Harrison 1999; Keller 2003; Smarzynska 2003). Indeed, in a review of 40 studies of horizontal productivity spill-overs spanning developed, developing, and transition economies, a World Bank paper found that only 19 reported positive, statistically significant spill-over effects (Gorg and Greenaway 2004). Moreover, studies using cross-sectional data are likely to be biased because of two data problems (ibid., p. 7). First, they use data at one point rather than over time. Second, they fail to control for other factors that might affect differences in productivity between industries. As a result, what industry-level studies demonstrate is not cause-and-effect but correlation. In other words, rather than spill-overs from FDI to domestic firms, higher average productivity in an industry with a high concentration of FDI may simply indicate that MNCs were attracted to that sector because productivity was already relatively high. For example, if productivity in the steel industry is higher than in metals processing, FDI is likely to be drawn more to the former than the latter. "In a cross section," Gorg and Greenaway point out, "one would find a positive and statistically significant relationship between the level of foreign investment and productivity, consistent with spill-overs, even though foreign investment did not cause high levels of productivity but rather was attracted by them." To correct for this, researchers must use firm-level panel data. Returning to their survey and eliminating the biased cross-sectional studies, Gorg and Greenaway found that only 6 of the 40 studies found evidence of positive spill-overs. *None* of the positive spill-overs were in developing countries.

Moreover, 6 of the 28 studies of developing and transition economies found evidence of *negative* spill-overs. In the study quoted most often, Aitken and Harrison (1999) analyzed a panel of more than 4,000 manufacturing firms in Venezuela between 1976 and 1989. They found that the total factor productivity of local firms in the same industry *decreased* in the presence of MNCs.[8]

Another line of investigation considers whether FDI differentially affects domestic firms depending on their "absorptive capacity." In a study of MNCs in Argentina's manufacturing sector in the 1990s, Chudnovsky et al. (2004) found that, on average, FDI had neither positive nor negative spill-overs.[9] However, some domestic firms reaped positive spill-overs, improving market position, while others suffered negative spill-overs and went out of business. The effect extended both to competitors and downstream suppliers of MNCs.

Case studies present a more optimistic, though still mixed, assessment of the potential for FDI to generate intra-industry spill-overs. Amsden and Chu (2003) found that partnerships with MNCs in the 1960s and the 1970s helped build globally competitive electronics firms in Taiwan. On the other hand, Barclay (2003) found that substantial inflows of FDI in the 1990s failed to stimulate domestic development in the natural gas industry in Trinidad and Tobago (Barclay 2003). Ernst (2003, p. 4) found that, despite large MNC investment in its electronics sector, "Malaysia has failed to develop a sufficiently diversified and deep industrial structure to induce a critical mass of corporate investment in specialized skills and innovative capabilities."

Vertical Spill-Overs

Some analysts argue that the lack of evidence that MNCs generate spill-overs to domestic firms in developing countries suggests that researchers "have been looking . . . in the wrong places" (Smarzynska 2003, p. 2). Because MNCs seek to keep knowledge from leaking to competitors but want and need to transfer it to local suppliers, spill-overs are more likely to be vertical than horizontal (Saagi 2002). The place to look is thus in the domestic firms who supply MNCs. A study of FDI in Lithuania, for example, found evidence of positive vertical spill-overs in the manufacturing sector (Smarzynska 2003).

Case studies, too, provide some support for vertical spill-overs. Moran (1998) found that FDI by the "big three" US car companies in the 1980s worked to upgrade technology and global competitiveness of Mexican auto supply firms. Singapore heavily depended on MNC investment to develop a dense network of local supply firms in its electronics sector in the 1970s and the 1980s (ibid.; Wong 2003). Other case studies, however, have found that MNCs generate few backward linkages to local firms. In northern Mexico, Brannon et al. (1994) found that, despite 25 years of FDI, Mexican material inputs accounted for less than 2 percent of value added in maquiladora plants. Based on surveys with plant managers and corporate purchasing agents, the study found that MNCs' purchasing strategies favored imports over domestic firms (Brannon et al. 1994).

The "Crowding In" of Domestic Investment

Investment is a necessary condition for economic growth. Although concepts of specific linkage mechanisms vary, virtually every economic theory highlights the importance for growth of net additions to physical (produced), natural, and/or human-capital assets.[10]

Beyond simply adding to capital formation, the promise of FDI is that it "crowds in" domestic investment. "Crowding in" occurs when domestic investment takes place in the presence of MNCs that would not have occurred in their absence. "Crowding in" is thus a macroeconomic externality of FDI. In terms of the national account, total gross fixed capital formation is the sum of foreign and domestic investment in physical capital. If a dollar of FDI adds a dollar to total capital formation and domestic investment remains unchanged, the "crowding" effect is neutral and there is no externality. "Crowding in" occurs when a dollar of FDI adds *more* than a dollar to total investment by stimulating an increase in domestic investment. "Crowding out" is the opposite: a dollar of FDI reduces total capital formation by diminishing domestic investment.

FDI can "crowd in" domestic investment in four ways:

Backward linkages   Local firms supply production inputs to MNCs.

Forward linkages   Goods and services produced by MNCs are efficiency-enhancing inputs to other industries.

Knowledge spill-overs   Domestic firms became more productive because of the diffusion of MNC technology and tacit knowledge.

Multiplier effects   Increased employment in MNCs increases local spending on domestically produced goods and services.

While necessary, one or more of these linkages from MNCs to domestic firms may not be sufficient to "crowd in" new investment. Firms must have two additional capabilities: access to credit and entrepreneurial skills. "Crowding in" is more likely in countries that offer well-functioning domestic financial markets and opportunities for business management training.

FDI crowds out domestic investment in a variety of ways. MNCs can out-compete domestic firms either because they are more efficient or because, as oligopolies, they undercut domestic prices and have access to global markets. MNCs also have access to internal finance and global capital markets, while domestic firms usually must borrow in domestic financial markets. If MNCs do borrow on domestic financial markets, they may crowd out local firms by driving up interest rates.[11] MNCs may also generate negative knowledge spill-overs, for example, by hiring domestic entrepreneurs and skilled workers away from domestic firms.

Whether FDI crowds domestic investment in or out is an empirical matter. Studies in developing countries have found evidence of both. For example, a study by the Brookings Institution covering 58 countries in Latin America, Asia, and Africa between 1978 and 1995 found that a dollar of FDI generated another dollar in domestic investment (Bosworth 1999). Renowned Indian economist Nagesh Kumar found the opposite. Based on data from 55 developing countries between 1980 and 1999, he concluded that "in net terms, the effect of FDI on domestic investments appears to be negative" (Kumar 2002). However, for a few countries, FDI crowded in domestic investment.

The Chilean economists Manuel Agosin and Ricardo Mayer found that the impact of FDI on domestic investment differed markedly by region and, to a lesser extent, among countries within a region. Using panel data for the period 1970–1996, they found that "in Asia—but less so in Africa—there has been strong "crowding in" of domestic investment by FDI; by contrast, strong crowding out has been the norm in Latin America" (Agosin and Mayer 2000, p. 1). Indeed, while FDI inflows into Latin America in the 1990s were 13 times what they were in the 1970s, GDP growth was 50 percent lower (Gonzalez 1999).

What accounts for the heterogeneity of the impact of FDI in different countries, especially the great difference in the experience Latin America versus Asia? "Little can be said on an *a priori* basis," Agosin and Mayer suggest (2000, p. 3). "The effects of FDI on investment may well vary from country to country, depending on domestic policy, the kinds of FDI that a country receives, and the strength of domestic enterprises."

One hypothesis is that it depends on which sector of the economy the FDI goes into. There are two opposing variants of the argument. On the one hand, some analysts argue, FDI flows into a relatively undeveloped sector are more likely to "crowd in" new investment because FDI brings technology, skills, and global market linkages that were absent from the host economy. The new assets generate knowledge spill-overs and catalyze growth. By the same token, FDI into more developed sectors has a greater propensity to substitute for local firms and crowd out domestic investment. For example, big MNC retailers like Wal-Mart and Telco are likely to crowd out domestic mom-and-pop stores.

On the other hand, it may be that in a less developed sector, for example high technology, FDI could pre-empt the emergence of domestic firms. East Asian countries such as Taiwan and Korea limited the presence of MNCs in the high-tech sector for just this reason. However, East Asian governments were particularly effective in implementing policies to nurture domestic firms, including in the way they utilized FDI (Amsden 2001). Few developing countries today have similar state capacities.

Another possibility is that FDI is more likely to "crowd in" investment when it adds new production capacity (a greenfield) and to crowd out investment when it involves simply the transfer of asset ownership. However, studies in Latin America suggest that, in some cases, MNCs made substantial post-purchase investment in modernization and rationalization, including technological upgrading (Agosin 1996). In other cases, however, MNCs acquired domestic companies that were already using state-of-the-art technology and neither made nor stimulated additional investment. In these cases, FDI was more akin to a portfolio investment than a true addition to capital formation in the national-accounts sense (Agosin and Mayer 2000).

The potential for FDI to "crowd in" domestic investment may also depend on generally strong macroeconomic conditions, as reflected by an already high rate of domestic investment. In Latin America, the rate of domestic investment has been much weaker than in Asia. "It could also be," Agosin and Mayer conclude, "that Latin American countries have been much less choosy about FDI than Asian countries, either in the sense of prior screening or attempting to attract desirable firms" (2000, p. 15).

In addition to being choosy about their MNC partners, East Asian governments proactively articulated what they wanted from FDI, including targeted backward and forward linkages and knowledge spill-overs. They were also able to effectively design and implement policies that met their objectives, including selecting sectors open to FDI and actively seeking partnerships with particular MNCs (Amsden 2003; Lall 2004). In Latin America, governments adopted a much more laissez-faire and passive approach that aimed to maximize the quantity rather than the quality of FDI inflows.

Sustainable-Development Spill-Overs

A central promise of FDI is that it will promote not just economic growth but more *environmentally sustainable* growth in developing countries.[12]

"FDI has the potential to bring social and environmental benefits to host economies," argues the OECD (2002), "through the dissemination of good practices and technologies within MNCs and through their subsequent spill-overs to domestic enterprises." While it cautions that foreign-owned enterprises may instead export old technologies and "bad" practices banned in host countries, the OECD optimistically concludes that "there is little empirical evidence to support the risk scenario" (ibid.).

Many studies in the 1990s examined whether FDI generates "pollution havens" in developing countries. Generally, they found that, for most industries, environmental management constitutes too small a percentage of total production costs to affect MNCs' location decisions (Jaffe 1995; Eskeland and Harrison 1997).[13] More important than what draws a multinational corporation to a particular country, however, is how it performs once it gets there. More than faith, a leap from the "risk scenario" to a "benefit scenario" requires both analytical rigor and empirical review.

Analytical Framework
FDI potentially delivers three types of spill-overs for sustainable industrial development:

- clean technology transfer (transfer to MNC affiliates of production technologies that are less polluting and more input-efficient than those used by domestic firms)
- technology leapfrogging (transfers of state-of-the-art production and pollution-control technologies by MNCs that allow developing countries to leap to the global technology frontier)
- pollution halo (transfers to MNC affiliates and diffusion among domestic firms, including suppliers, of "best practice" in environmental management).

The existence of these spill-overs in developing countries rests on four assumptions. First, because they are subjected to global competition, MNCs are more technologically dynamic than domestic firms. Technological change tends to promote efficiency in the use of production inputs, which reduces both pollution and resource intensity. Second, MNCs transfer their cleanest, state-of-the-art production technologies to developing countries. They do so because developing-country production sites are integrated into MNCs' global production and marketing strategies. Competing for global markets requires companies to maximize efficiency at all production sites. Third, originating primarily in OECD countries, MNCs are subject to higher environmental standards in their home countries than are domestic firms in developing countries. Reflected in both government regulation and consumer preferences, higher environmental standards force MNCs to direct R&D funds toward cleaner and safer process technologies and products. They also nudge MNCs "beyond compliance" with mandatory environmental regulation and toward adopting best practice in environmental management as a way of demonstrating corporate social responsibility. Increasingly, best practice includes corporate oversight not only of foreign-based subsidiaries but also of supply chains. Fourth, to minimize transaction and reputation costs and liability risks, MNCs operate with global, company-wide environmental standards. These centralized standards are based either on those of the MNC's home country or on higher internal or international standards.[14] It is "standard operating procedure" for MNCs to diffuse these standards to foreign affiliates and monitor performance. Where these assumptions hold, it is reasonable to expect that FDI will generate sustainable-development spill-overs. However, such assumptions may not hold for all or even most MNCs.

In terms of technological dynamism, industries and industry subsectors are subjected to differing degrees of global competition and other drivers of technological change. The mining industry, for example, is far less dynamic in terms of innovation in process technology and final products than the informatics industry. Even within a relatively dynamic industry, companies are not homogeneous: there are leaders and laggards in terms of both technological and managerial innovation, including in environmental management (Mazurek 1999; Leighton et al. 2002). Whether FDI generates sustainable-development spill-overs is thus likely to be contingent on the technological trajectory within a particular industry, as well as the innovation culture of particular companies.

Like knowledge spill-overs more generally, MNCs have incentives to protect intellectual property in environmentally cleaner technology and/or better management practices. Even in technologically dynamic industries and companies, MNCs may find it cost-effective to slough off older, lower-margin, dirtier technologies to developing countries rather than transfer state-of-the-art production technologies.

The assumption that environmental regulation of industry generally is more stringent in OECD than in developing countries is probably warranted. However, though they tend to converge, government regulation and social expectations are not uniform in OECD countries. The European Union, in particular, has in recent years adopted more stringent standards for industry, including regulation of industrial chemicals, than North America. European MNCs have also voluntarily adopted "best practice" at a higher rate than those in North America, Australasia, or Japan.[15] In addition, MNCs increasingly are emerging from non-OECD countries, including China, Singapore, and Malaysia (UNCTAD 2004). Environmental practices and the influence of civil society are not only generally lower but more diverse than in developed countries. In short, the environmental management practices that MNCs bring to developing countries are conditioned by their source country and thus are highly diverse.

Finally, the assumption that MNCs uniformly adopt and implement global standards as a strategy for cross-border environmental management is too optimistic. In many developing countries, environmental regulation is non-existent, weak or not enforced. In this context, foreign firms have a choice of four strategies: (1) follow (or set) local practice, (2) comply with host-country regulation, (3) follow home-country standards, (4) operate with higher standards set by the company or by an international agency (Hansen 1999). Which strategy they choose depends on their determination of costs and risks.

Trade-offs between costs and risks are complex and conditioned by multiple factors, including global market dynamics within a particular sector and the size and global strategy of particular MNCs. On the one hand, the adoption of global standards reduces transaction and information costs entailed in producing to a myriad of different national environmental and occupational health standards. Some MNCs operate in dozens of countries. High company-wide standards may also reduce production costs by increasing eco-efficiency, that is, reducing inputs per unit output and reducing pollution. Moreover, high company-wide standards reduce the risk of liability and adverse publicity resulting from high-profile environmental disasters and/or targeted activist campaigns.

On the other hand, two factors militate against multinational corporations' adoption of high company-wide standards, whether based on home-country, internally set, or international standards. First, MNC liability for environmental harm in developing countries is generally limited to compliance with domestic law. In international law, the concept of "environmental rights" with the right of redress is generally limited to governments and to cross-border environmental harms. New forms of liability, including of financial institutions that make loans to MNCs and home-country governments, are still in an early stage.[16] Second, reputation costs are negligible for MNCs without high-visibility consumer products and brands. Even for those MNCs vulnerable to loss of consumer confidence, the probability of suffering reputation costs due to poor environmental practice may be low because of poor monitoring, reporting, and information mechanisms. Put simply, they have a low risk of getting caught or being held to account if they are caught engaging in "bad practice," not least because of the limited resources of advocacy groups who are the information bridge to consumers.

Examining the Evidence
Analytical arguments both support and challenge the idea that FDI generates sustainable-development spill-overs in developing countries. The empirical evidence likewise paints an ambiguous picture.

Are foreign firms cleaner than domestic firms?
A detailed case study of FDI in Chile's mining sector in the 1970s and the 1980s found that the two foreign-owned companies performed far better than domestic companies. Following their corporate parents, Exxon Minerals Chile and Fluor Corporation, the two foreign-owned companies put in place an environmental policy framework requiring "responsible practices" at a time when there was as yet no coherent government regulation of the mining industry (Lagos 1999). Ten years later, domestic mining companies followed suit.

Using survey methodology, a study of MNCs in India's manufacturing sector likewise found that foreign firms were cleaner than domestic firms

(Ruud 2002). MNCs transferred state-of-the-art production (though not necessarily pollution control) technologies to their affiliates. In addition, MNC affiliates were strongly influenced by corporate parents to improve environmental management.

Other case studies, however, suggest that MNCs follow poor local practice or, in cases where domestic companies do not exit, engage in "bad practice." In explorations for oil and natural gas off Sakhalin Island, for example, Exxon openly flaunted new Russian laws requiring environmental review and zero water discharge. Only the pressure of environmental jurists and activists, as well as the European Bank for Reconstruction and Development, forced Exxon to comply with the law (Rosenthal 2002). Case studies of FDI in the petroleum industry, including in Nigeria, Ecuador, Azerbaijan, and Kazakhstan, likewise find that MNCs operate in developing countries with "double standards"—environmental and human rights practices that would be fined or prosecuted in their home countries (Leighton et al. 2002).

Statistical studies are likewise mixed. Using energy use per unit of output as a proxy for energy emissions, one World Bank study found that foreign ownership was associated with cleaner and lower levels of energy use in Mexico, Venezuela, and Cote d'Ivoire (Eskeland 1997). In China, foreign investment in electricity generation was linked to improvements in energy efficiency and emission reduction (Blackman 1998). Besides transferring advanced generating technologies and better management, FDI stimulated competition among Chinese companies in the electricity-generation sector.

Another group of World Bank researchers, however, found that foreign firms and plants performed no better than domestic companies. Based on firm-level data in Mexico (manufacturing) and Asia (pulp and paper), the New Ideas in Pollution Regulation group found firm environmental performance to depend on (1) the scale of the plant (bigger is better) and (2) the strength of local regulation, both government and "informal" (Hettige 1996; Dasgupta et al. 1997). In addition, a study of the manufacturing sector in Korea found that domestic firms performed *better* than foreign-owned firms, a result the authors attributed to the sensitivity of Korean *chaebol* to public criticism.

Does FDI promote technology leapfrogging?

Technology gaps between developed and developing countries can be very large. In this context, MNCs can transfer technology that is cleaner than that currently in use in developing countries, yet which is not state-of the-art. Moreover, production processes, especially in manufacturing, are complex and multi-stage. MNCs may transfer a mixture of older and state-of-the art technologies.

The concept of technology leapfrogging is that, through transfer by MNCs of the most efficient, least polluting technologies, developing countries can move to the global production frontier. The economic benefits include developing globally competitive industries, potentially even out-competing producers with older technologies. The reduction in the pollution-intensity of developing-country production (and consumption) brings environmental benefits both locally and (in the case of global pollutants such as carbon emissions) globally.

Three studies of the Mexican steel industry found that foreign firms were cleaner than domestic firms and that FDI generates environmentally beneficial technology leapfrogging. Mercado (2000) found that foreign firms in the Mexican steel sector, or firms that serve foreign markets, were more apt than domestic firms to comply with environmental regulations. Gentry and Fernandez (1998) found that the Mexican government played an important part, brokering an early agreement with Dutch steel firms whereby the government took on some of the environmental liabilities associated with steel production. Later, the foreign firms began investing in environmental improvements. Most tellingly, Gallagher (2004) found that steel production in Mexico is "cleaner" per unit of output, in terms of criteria air pollutants, than in the United States.[17] The primary reason is that FDI, as well as domestic investment in new plants, deployed newer and more environmentally benign mini-mill technology rather than more traditional and dirtier blast furnaces (ibid.).

But a recent study of FDI by US auto companies in China had less optimistic findings (K. S. Gallagher 2006). Based on extensive interviews with plant managers at Ford, GM, and Jeep affiliates, the study found that US firms transferred outdated automotive pollution-control technologies. While "somewhat cleaner" automotive technologies were transferred, potential environmental benefits will be outweighed by the increase in the number of vehicles. Most important, FDI did little to improve Chinese technological capabilities because "US companies have transferred products, but not much knowledge, to China" (ibid., p. 9). As a result, China (and the world) will reap neither the environmental nor the economic benefit of producing at the global technology frontier.

In earlier work, Gallagher (2004) posits that there are "limits to leapfrogging" through technology transfer from foreign firms via FDI. In the case of substantially cleaner automotive technologies in China, these limits included, first and foremost, the lack of strategic and consistent Chinese government policies that aimed to achieve environmental and energy-technology leapfrogging.[18] In addition, China had weak technological capabilities, and MNCs were unwilling to transfer cleaner or more efficient technologies beyond those required by standards. Gallagher concludes that, for China to leapfrog to the technological frontier of clean automobiles, "[a] coherent, concerted, consistent, and long-term effort of government, industry, and civil society cooperation" would be required (ibid., p. 10).

Do MNCs diffuse good environmental management to domestic firms?
Beyond transferring clean technology and good environmental management to their affiliates, multination corporations can generate sustainable-development spill-overs by diffusing good environmental management to domestic firms. One of the primary channels is through the supply chain; that is, MNC requirements that suppliers meet their internally set environmental standards. Environmental management training opportunities for suppliers accelerates diffusion. Another channel is the demonstration effect. Domestic firms copy foreign firms or host-country governments adopt MNC standards as local regulation. A third channel is industry collaboration to promote better environmental management in developing countries through self-regulation.

A study of MNCs in the chemical industry in Latin America found that US companies played a leading role in diffusing the Responsible Care program to domestic companies in Mexico and Brazil (Garcia-Johnson 2000). Developed in the early 1980s in Canada by the chemical industry in response to the Bhopal disaster, the program aims to raise industry self-regulation beyond mandatory government standards in the areas of environmental impact, employee health and safety, facility security, and product stewardship.[19] By 2005, chemical industry associations in 45 countries had signed up to the Responsible Care program, including 17 in developing countries.[20]

In a volume of case studies in Latin America, Gentry (1998) found evidence that better environmental management practices were diffused through FDI, including through supply chains. In India's manufacturing sector, however, Ruud (2002) found no evidence that MNCs diffused better environmental management practices to local partners, suppliers or consumers. While MNC affiliates were cleaner then domestic firms, they apparently operate as "islands of environmental excellence in a sea of dirt" (ibid.). Ruud concludes that local norms and institutions are central in determining MNC practice and that "FDI inflows do not automatically create a general improvement in environmental performance" (ibid., p. 116).

From Ideology to Problem Solving

With its foundation in neo-classical economic theory, the mainstream development paradigm advises that "letting the market do it" is the best way to promote sustainable economic growth in developing countries. The role of government is to provide unfettered access, protect contracts, keep the social peace, and remove policy restraints. MNCs will then integrate developing countries into their global sourcing and marketing strategies, nurturing the growth of domestic firms through spill-overs.

Evidence that FDI generates knowledge, investment or sustainable-development spill-overs in developing countries, however, is thin and, at best, ambiguous. Statistical studies have found *no* evidence that FDI generates positive knowledge spill-overs—and some evidence of negative spill-overs—to developing-country firms in the same industry. Things look a bit better for downstream, supplier firms, though there are only a few studies to date. Case studies, on the other hand, have shown that FDI can deliver positive or negative spill-overs to domestic firms in both the same and upstream industries. The evidence likewise suggests that FDI might or might not "crowd in" domestic investment. Why the gap between theory and evidence?

The problem may stem in part from research methods. More consistent use of firm-level panel data may allow statistical studies to yield more consistent results and new insights. On the other hand, case studies suggest that there simply is no consistent or guaranteed relationship between FDI and spill-overs because the crucial determinants—including company and government policy—are contingent. As a whole, existing studies point toward the need to revisit and refine the theory in three ways.

Spill-Overs May Not Exist
The failure for FDI to generate knowledge spill-overs may lie on the "supply side": there may be little leaking out. MNCs tend to actively work to not leak proprietary knowledge, including through strategies such as employee-retention incentive programs and keeping host-country nationals out of senior management positions. In addition, the global production, sourcing and marketing strategies of MNCs may reduce the likelihood that there is much knowledge to leak in developing countries. For example, if investment is motivated by the local availability of cheap labor, MNCs are likely to transfer their simplest and least skill-intensive technology. Likewise, if the MNC motivation is to gain a monopolistic position by buying up existing firms, domestic market structure will become less rather than more efficient and competitive (Soreide 2001).

Moreover, MNCs generally retain their most knowledge-intensive function—research and development—in their home countries.[21] "Manufacturing MNCs," Kim argues (2003, p. 146), "set up foreign plants to optimize their global sourcing of inputs and production of outputs. To do this, they transfer the production and management capabilities needed to ensure efficient production. Some MNCs also undertake R&D in host countries, but this is mainly in order to adapt products to local or regional needs. Very rarely do they transfer advanced engineering and innovation capabilities."

Global MNC sourcing strategies are particularly important to developing countries, given the greater likelihood of capturing vertical spill-overs. In many industries, however, MNCs are biased toward global suppliers, reducing the potential for vertical spill-overs through backward linkages.

The propensity to both generate and share knowledge differs between and within sectors and industries. Some industries, such as information and communications technologies, are highly dynamic in terms of generating both new products and production technologies. Others, such as apparel and mining, tend to utilize stable technologies to produce fairly non-differentiated products. Indeed, one study found the effect of FDI on economic growth to differ markedly by sector: in the primary sector, FDI reduced economic growth while in the manufacturing sector, FDI increased growth. The effect in the service sector was ambiguous (Alfaro 2002).

The potential for knowledge to spill over may also vary by firm. Within an industry, some firms are technological leaders, while others are laggards. Some firms are also normative leaders, articulating environmental and human rights policies as part of their "corporate social responsibility" and building the capacities of local suppliers to implement them. There are also differences between foreign subsidiaries within a company. Some MNC affiliates have created substantial "knowledge creating and accumulating activities" of their own; others have not (Marin 2003).

Spill-Overs Must Be Captured

Some studies found evidence of positive spill-overs, suggesting that their elusiveness may be due less to problems on the supply than on the "demand side." MNCs may leak or even intentionally seek to transfer knowledge in developing countries—but domestic firms are not able to capture and commercialize it. Explanations for this failure revolve around the concept of "absorptive capacity"—the capability to integrate and exploit external knowledge in processes production and innovation. Originally analyzed at the level of the firm (Cohen 1990), absorptive capacity also refers to host-country characteristics.

For firms, absorptive capacity is the stock of pre-existing skill and experience in integrating new knowledge. Gauges of differential capacities to absorb and commercialize knowledge include skill intensity (ratio of skilled to total workforce); spending on innovation activities, including R&D, per employee; investment in capital goods per employee; and ratio of exports to total sales (Chudnovsky et al. 2004).

Some studies suggest that what crucially determines whether spill-overs are captured is the size of the technological gap between multinational and domestic firms. Findlay (1978) finds that the bigger the gap, the greater the likelihood of capturing spill-overs, because of the greater potential for learning and leapfrogging. More recent and studies argue persuasively that, because a large gap requires large investment costs to catch up, the likelihood of capturing spill-overs is greater with a smaller gap (Sawada 2004).

At the national level, determinants of absorptive capacity include the quality of "human capital" (that is, the skill, literacy level and discipline of the work force) as well as the quality of local infrastructure such as transport and communication. Several studies (e.g., Borensztein et al. 1998) have found the role of FDI in economic growth to be positively correlated with the level of workforce education and infrastructure.

Another national capacity factor is the state of local financial and credit markets. Most domestic firms seeking to commercialize a new product or invest in new technology have access only to local finance. If credit is not available or the cost of capital is high, opportunities will be lost. One study found that countries with well-developed financial markets "gained significantly" from FDI, while those without them did not (Alfaro 2002).

The Host Country's Policies Are Pivotal
Both case studies and several statistical studies point to the overarching conclusion that the host country's policies are pivotal in gaining positive spill-overs from FDI. National policies influence both the supply of knowledge spill-overs by influencing MNC global strategies, and the potential for domestic firms to capture spill-overs by enhancing absorptive capacities.

A number of studies found spill-overs to be contingent on local factors, including the size of the technology or productivity gap between the MNCs and domestic firms (Kokko et al. 1996). A review of the literature found the impact of FDI "varies between industries and countries depending on country characteristics and the policy environment" (Blomstrom and Kokko 1996).

What constitutes a "conducive" policy regime? There is general agreement about the importance of government in enhancing national absorptive capacity by investing in primary, secondary, and tertiary education and in worker training, as well as by improving the functioning of domestic financial markets. Some researchers also point to government policies that strengthen the "national innovation system," that is, the overall capacity to learn and exploit knowledge, including through science and technology policy (Lall 2004).

There is lively debate, however, as to whether government should play a more targeted interventionist role to capture spill-overs for domestic firms or, on the contrary, reduce obstacles to MNC global production, marketing and sourcing strategies. Interventionist policies include, *inter alia,* targeted subsidies to domestic firms and performance requirements for MNCs, such as making access to the domestic market contingent on exporting a certain proportion of products, requiring MNC training programs as a condition of market access, and encouraging backward linkages through domestic-content requirements.

Despite his own findings that selective industry policies spurred MNCs to upgrade Mexico's auto-parts industry, Moran (1998) argues that a liberal trade and investment regime allowing MNCs maximum flexibility has the best chance of increasing the efficiency of local firms by integrating them into global supply chains. On the other hand, Ernst (2003, p. 4) admonishes that "Malaysia's experience in the electronics industry indicates that nothing is automatic about benefits from participating in [global production networks]."

According to Amsden and Chu (2003), the most important ingredient in capturing spill-overs, and indeed in increasing productive capacity in "latecomer" states, is a strong state acting to nurture domestic firms through effective, market-friendly, and performance-related subsidies.[22] Barclay (2003, p. 1), who studied the oil and gas industry in Trinidad and Tobago, concludes: "FDI-facilitated development is not an effortless process. It only occurs when host developing-country governments implement selective intervention policies that are aimed at increasing indigenous technological capabilities."

Regardless of their efficacy, many selective intervention policies that were effective in the past are now prohibited by trade and investment rules, such as the Trade Related Investment Measures agreement of the World Trade Organization. One important question is how much "room to move" governments now have to design policies and programs that will encourage MNCs to transfer knowledge and will enhance the capacities of domestic firms to capture it. Case studies (Evans 1995; Lall 2004) suggest that there may be plenty of "room to move." The first step—to articulate a proactive development strategy rather than rely simplistically on FDI—might be the most important. In countries where FDI generated knowledge spill-overs, including Singapore, Taiwan, Korea, and Brazil, governments were able to set objectives in terms of what they wanted to gain from FDI, and to implement a coherent set of policies to gain their objectives. Innovative policies to build local institutional capacities were implemented in the areas of R&D, tertiary education, and science and technology. Governments also brokered relationships between universities and business, including incentives for MNC training programs for local suppliers. All these initiatives are still permitted by the WTO and under various trade agreements.

The Need to Be Proactive

Taken as a whole, the evidence suggests that the orthodoxy about unfettered FDI as a driver of global development is wrong. On the one hand, many middle-income developing countries, especially in Latin America, have attracted FDI inflows but have not captured spill-overs and increased growth. On the other hand, most countries that successfully utilized FDI to bolster domestic productive capacities, especially in East Asia, did so by implementing strong and somewhat restrictive government policies.

The "let markets do it" ideology is too homogeneous and simplistic. Developing countries (and locales within countries) are highly heterogeneous, differing not only in terms of income, size, technology, skill base, and culture but also in their development policies and regulatory institutions. All these factors affect the local "absorptive capacity" for FDI.

Moreover, FDI is itself highly heterogeneous, spanning all sectors—primary, manufacturing, and services—and encompassing both additions to investment (greenfield) and the purchase of existing firms (mergers and acquisitions). Both the sector and the type differentially affect the development benefits of FDI. Moreover, there is also firm-level heterogeneity within sectors and industries. MNCs have different global procurement, marketing, and investment strategies, as well as different technology trajectories and management cultures. The capture of economic benefits from FDI is contingent on both the industry and the particular firm (or firms).

Rather than simply removing fetters and increasing protections to foreign investors, the capture of positive benefits from FDI depends on a kind of "supply-and-demand" partnership between MNCs and governments—that is, charting and implementing a negotiated intersection of the global strategy of a particular MNC with local objectives, policies, and institutions.[23] Different models of a host country's interaction with MNCs have yielded different results for sustainable industrial development.

In general, an analysis of the literature points to the conclusion that host countries that derived substantial benefits from FDI, especially in East Asia, adopted a *targeted and proactive* model (Lall and Urata 2003). Governments strategically defined medium-term and long-term industry and sector-level objectives for domestic firms and selectively attracted MNCs that could offer targeted technology and skills. They also imposed a variety of performance requirements on MNCs, including requirements to establish joint ventures with domestic firms, to produce for export, to provide training to local personnel, and to use domestic content. More recently, they have aggressively utilized education and training, science and technology policy, and public-private partnerships with MNCs to promote further industrial evolution.

Countries that have been less successful in gaining benefits from FDI, including in Latin America, adopted a *diffuse and passive* model focused more on the quantity rather than the quality of FDI inflows. They eschewed industrial policy, liberalized investment, and concentrated on FDI inflows primarily as an engine of employment, rather than as a way to boost the targeted growth of domestic firms.

Grand claims aside, MNCs are highly heterogeneous in their willingness to transfer technology and know-how to developing countries and their environmental management practice. Governments, too, are highly heterogeneous in terms of their local institutional capacities and development objectives. Whether and how FDI crowds in investment or generates knowledge and sustainable-development spill-overs are matters to be discovered not in theory but in practical, problem-solving approaches to the design of policy and negotiations with MNCs.

2

The Emergence of Mexico's Enclave Economy

Mexico's liberalization strategy of the 1990s aimed to stimulate domestic economic growth by increasing the productivity and competitiveness of export-oriented manufacturing. Eschewing past industry and macroeconomic policies that promoted domestic firms, liberalization policies favored foreign firms. While industry policies were mostly "neutral," macroeconomic policies, especially high interest rates and an overvalued exchange rate, created a climate conducive to foreign but problematic for domestic investment. The hope was that the benefits of foreign investment would "spill over" to local firms, boosting domestic productivity, employment, and economic growth.

There were also hopes that FDI-led growth would bring environmental and social benefits. The growth of manufacturing jobs would absorb the urban poor and farmers displaced by NAFTA, closing the income inequality that plagues Mexico and stemming rural-urban and cross-border migration The more efficient, globally integrated foreign firms would transfer "clean technology" and systems for better environmental management, reducing the pollution and health risks associated with industrial development.

By one set of standards, the strategy was successful. During the 1990s inflows of FDI increased by a factor of 5 over the inflows of the 1980s, and about half of that FDI flowed into the manufacturing sector. Exports increased by more than a factor of 3, and manufactures accounted for nearly 90 percent of the total. However, a current-account deficit has persisted. Moreover, FDI inflows fell sharply after 2001, due largely to the downturn in the US economy and China's accession to the WTO.

A simple perception of success, however, obscures a confusion of means and ends. The central goal of a development strategy is—or should be—not to increase FDI and exports but to improve the lives of people, both in the short and longer term. Central to achieving this aim is the growth of sustainable industry.

This chapter examines the recent performance of the FDI-led development strategy against the goals of sustainable industrial development.

From Import Substitution to Foreign Investment

Mexico has tended to swing like a pendulum in response to economic crises. The first swing, toward import-substituting industrialization (ISI), followed the Great Depression and World War II. By 1980, Mexico was the flagship of ISI policies in Spanish America. In the early 1980s, Mexico began a swing in the other direction, toward economic openness and integration. By the turn of the century Mexico had joined the General Agreement on Tariffs and Trade and the Organization for Economic Cooperation and Development, had negotiated the North American Free Trade Agreement, and was a strong advocate for global trade liberalization at the World Trade Organization.

When World War II came to an end, the gap in per-capita income between the developed and the developing countries became an area of grave concern in the developing countries. In response, the priority for developing countries became raising national incomes. Rather than relying on foreign capital and global markets by liberalizing their trade and investment regimes, many developing countries, including Mexico, chose to chart a path toward development that promoted industrial self-sufficiency and growth in domestic markets. ISI performed remarkably well in Mexico and other Latin American countries for more than 30 years.

Many economists argued that developing economies should restructure industry away from agricultural and extractive sectors for export to manufacturing for both domestic and export markets. In Mexico and elsewhere, the tools of ISI focused on a number of key policies, including major public outlays for infrastructure, planning, tariffs, import licensing, quotas, exchange rate controls, wage controls, and direct government investment in key sectors (Cardoso and Helwege 1993; Bruton 1998). From the beginning of World War II until the early 1970s, this strategy performed well in Mexico. Indeed, this period is often referred to as Mexico's "Golden Age." During this time, the economy grew at an annual rate of more than 6 percent (more than 3 percent per capita). The engine of growth was the development of a strong manufacturing sector.

Manufacturing growth in Mexico was a function of a developmental state. Mexico industrialized through building public infrastructure, conditional government support, and import substitution (Moreno and Ros 1994). Government subsidies and import protection, in addition to loans from national development banks, were given to Mexican industry in exchange for concrete results, including local content requirements, price controls, technological innovation, capacity, and exports (Anderson 1963; Blair 1964). Through this process, Mexico created "national leaders" in the form of key state-owned enterprises (SOEs) in the petroleum, steel and other industries. These sectors were linked to chemical, machinery, transport and textiles industries that also received government patronage (Baer 1971; Amsden 2000). Indeed, in the first decades after World War II, these sectors received more than 60 percent of all investment, public and private (Aguayo Ayala 2000). By the 1960s, manufacturing was a large and growing share of total production in Mexico. In 1940, agriculture was 22 percent of total output and manufacturing was 17 percent (Reynolds 1970). By the early 1970s, agriculture had shrunk to around 10 percent while manufacturing had grown to nearly 23 percent.

In addition to state-owned enterprises and state-patronized private industries, Mexico established export-processing zones called maquiladoras in the mid 1960s. Maquiladoras are "in-bond" assembly factories where imports of unfinished goods enter Mexico duty-free provided that the importer posts a bond guaranteeing the export of the finished good. Many maquiladoras are located in the US-Mexico border region and include electrical and non-electrical machinery, much of the transport industry, and some apparel. The SOEs, the state-patronized private enterprises, and the maquiladoras supplied growing internal and external markets for their production.

By the late 1970s, Mexico seemed to be on the path to first world economic status. The discovery of massive amounts of Mexican oil in 1976 seemed to all but secure that path. From 1976 to 1980, total Mexican GDP grew by an annual average of more than 8 percent (table 2.1). Assuming that rapid growth would continue for years to come, Mexico's government and private sector embarked upon a binge of borrowing and public spending.

The borrowing binge, coupled with a fixed nominal exchange rate, generated a large external debt, as well as rising inflation, growing real-exchange-rate appreciation, and renewed current-account deficits (Kehoe 1995). From 1970 to the early 1980s, Mexico's foreign debt rose from $3.2 billion to more than $100 billion (Otero 1996). When oil prices suddenly dropped in 1982, a time of high world interest rates, Mexico announced that it was unable to meet its debt obligations—a "watershed event" for most developing countries (Rodrik 1999). A major devaluation plunged Mexico into economic crisis.

Between 1982 and 1985, Mexico tried and failed to respond to the crisis with another shot in the arm of the ISI model. The administration of Miguel de la Madrid (1982–1988) initiated the Program of Immediate Economic Reorganization (PIRE). First, the plan aimed to restore financial stability through peso devaluation and a cut in the government's deficit. In addition, the government adjusted the minimum wage and the wages of public employees to keep them below inflation.

On the trade front, the initial strategy was to further restrict trade. In 1982, tariffs were increased to 100 percent of the value of all imports,

Table 2.1
Selected annual growth rates, 1940–2000. Sources: Reynolds 1970; World Bank 2005.

	1940–1959	1960–1979	1985–2000
GDP	6.3	6.6	2.6
GDP per capita	3.2	3.4	0.8
Exports	4.5	8.0	10.6
Imports	6.8	5.9	12.8
Manufactures	7.7	6.6	3.7
Gross fixed capital formation	9.0	8.3	3.8
Services	n.a.[a]	6.6	2.5
Agriculture	5.0	0.9	1.0

a. Here and in other tables, n.a. means not available.

licenses were required for importing all goods, and foreigners were allowed no more than 49 percent ownership of Mexican enterprises. In addition, Mexico signed a loan agreement with the International Monetary Fund (IMF) for $3.7 billion, and borrowed $5 billion from commercial banks in the United States. Another $2 billion came from the Paris Club, an informal group of (mostly) European governments with large claims on other governments in the world economy (Lustig 1998).

The de la Madrid administration believed financial stability would be restored through a drastic reduction in the deficit and by a large devaluation of the peso. It was further predicted that these policies would reduce inflation and create a necessary trade surplus (Lustig 1992). Nonetheless, by 1985 Mexico faced a balance of payments crisis once again: fiscal discipline had swayed, IMF funding had ended, a massive and costly earthquake hit Mexico City, and oil prices started a sharp dive (Kehoe 1995). Once again, the value of the peso depreciated sharply and set the stage for a new experiment in Mexican economic policy (Ten Kate 1992).

Faced with the possibility of another crisis, de la Madrid nudged the pendulum in the other direction and experimented with neo-liberalism. He named his effort Apertura, signifying the "opening" of Mexico to foreign trade and investment. The de la Madrid government lowered the portion of imports subject to license requirements from 100 percent in 1983 to 35 percent in 1985 (General Agreement on Tariffs and Trade (GATT) 1993). During Apertura, tariff rates were also lowered. The maximum tariff rate in 1982 was 100 percent. By 1986, there were 11 tariff rates and the maximum rate fell to 45 percent. Further reductions became locked in when Mexico became a signatory to the General Agreement on Tariffs and Trade at the end of 1986.

New Goals, New Policies

While de la Madrid nudged Mexico gently toward economic openness, his successor, Carlos Salinas de Gortari (1988–1994) firmly shoved the pendulum toward neo-liberal integration.

The situation that Salinas inherited was a grave one. Yet another financial crisis had struck in 1987, creating hardship for both the poor and the rich. Millions who lived in extreme poverty found income and services falling, while many in the elite echelons of society watched their wealth evaporate. To Salinas, something drastic had to be done. The macroeconomic stabilization efforts of 1983–1985 and the trade policies of 1985–1987 had failed to generate economic reforms. This time, the Mexican government decided to bundle the two issues together into a grand strategy.

Embracing Globalization: NAFTA and Beyond

In Mexico's National Development Plan for 1989–1994, Salinas articulated three overarching goals: (1) achieving macroeconomic stability, (2) increasing foreign investment, and (3) modernizing the economy (Mexico 1989). The heart of the plan was in the manufacturing sector. By opening the economy and reducing the role of the state in economic affairs, Mexico would build a strong and internationally competitive manufacturing sector. Fueled by foreign investment, the development strategy would

• increase competitiveness and growth in the manufacturing sector by reducing the role of the state so markets would focus on low-wage comparative advantages that would lead to an increase in exports,

• increase foreign exchange earnings and FDI and thus provide the country with new international reserves and stability,

• upgrade the manufacturing sector with new technologies transferred by transnational corporations that would locate to Mexico,

and

• create new employment, thereby attracting workers from less efficient rural areas to manufacturing centers and explicitly providing a disincentive to leave the country for work in the United States.[1]

Meeting these goals required a top-to-bottom revamping of Mexico's foreign and domestic economic policies. Domestically, the government negotiated a series of economic "pacts." Signed by representatives of labor, agricultural producers, and the business sector, the pacts included cuts in the fiscal deficit, a further tightening of monetary policy in the form of higher interest rates to tame inflation and stabilize the exchange rate, further trade liberalization, and a commitment by industry not to raise prices and by trade unions not to press for wage increases above inflation (Lustig 1998). The policies related to the pacts reduced inflation from an annual rate of 159.2 percent in 1987 to 7.1 percent in 1994, while raising GDP by 23.1 percent. The new climate set the stage for a debt reduction agreement signed by Mexico and its foreign creditors, and for an increase in World Bank and IMF support (Lustig 1992).

Foreign economic policy centered on a further embrace of the Apertura policy and liberalization of trade and investment. Coupled with the pacts, these policies solidified Mexico's transition to neo-liberalism. Nowhere was this about face more evident than in Mexico's stance toward regional economic integration. In 1979, Ronald Reagan, campaigning for the Republican presidential nomination, proposed the negotiation of a trade accord among the North American nations. At that time, Mexico was stiffly opposed—an opposition that lasted through the 1980s (Barry 1995). However, after unsuccessful attempts to craft deals with Japan and Europe, it was Carlos Salinas who approached the Bush administration in 1990 about the possibility of a North American Free Trade Agreement (Winn 1992).

Negotiated during 1991 and 1992, NAFTA went into effect on January 1, 1994. All tariffs among Mexico, Canada, and the United States were to be phased out over a 15-year period, with most tariffs and quantitative restrictions lifted by 2004. In addition to the lifting of tariff restrictions, NAFTA also considerably liberalized investment (Wise 1998). The agreement also had "side" agreements on trade-related aspects of labor and environmental standards. On the heels of NAFTA, Mexico signed or committed to a flurry of bilateral and regional trade agreements. In 1994 alone, Mexico signed agreements with Costa Rica, Colombia, Venezuela, and Bolivia.

In addition to NAFTA, Mexico made unilateral changes in domestic regulations on foreign investment intended to prepare for and align with Mexico's new international commitments. In 1989, Mexico reformed its 1973 Law to Promote Mexican Investment and Regulate Foreign Investment by allowing 100 percent foreign ownership in many new sectors, making it quicker and easier to get the approval for new investment projects, and relaxing requirements tied to exports and local content quotas (UNCTC 1992). In 1993, these reforms were wrapped into Mexico's new Foreign Investment Law (Dussel 2000).

To make investments less cumbersome for foreign firms, Mexico also reformed its technology transfer requirements. Until 1973, Mexico's Technology Transfer Law was geared toward strengthening the bargaining positions of the recipients of foreign technology. All technology transfers had to be approved by the Ministry of Trade and Industrial Promotion, which monitored the extent to which technology transfer could be assimilated, generated employment, promoted research and development, increased energy efficiency, controlled pollution, and enhanced local spill-overs.

In 1990, the Salinas administration put forth a new technology transfer law relinquishing all government interference in the technology process to the parties involved in FDI. Government-enforced conditions on technology transfer were phased out, and technology agreements no longer needed government approval (but must be registered). Moreover, the law now contains strict confidentiality clauses (UNCTC 1992). These efforts were expanded upon and locked into place under NAFTA in 1994—at least pertaining to investment by the United States, the biggest investor in Mexico, as well as Canada. Under NAFTA (Article 1106), all performance requirements for foreign investors, including local content, export requirements, technology transfer, etc., were to be gradually eliminated by 2004 (Dussel 2000). However, in some sectors performance requirements were simply extended to the North American region as a whole (Moran 1998; Dussel 2003). For example:

• In the auto sector, 62.5 percent of automobile parts and components are required to be sourced from the NAFTA parties.

• Ninety percent of circuit board assemblies must be packaged in NAFTA countries.

• Photocopiers, printers, and fax machines must be sub-assembled in North America (seen as an equivalent to an 80 percent domestic content requirement).

• In order to qualify for preferential status, television tubes must be produced inside NAFTA countries.

In addition, NAFTA gives foreign investors the right to settle disputes through binding international arbitration for compensatory damages due to performance requirements and other forms of regulation that are deemed to be "tantamount to expropriation."

In the 1990s, Mexico became party to a number of new investment rules agreed to under the WTO, including Trade-Related Investment Measures (TRIMS) and Trade in Intellectual Property Rights (TRIPS). Both TRIMS and TRIMS limit the ability of Mexico to impose performance requirements on foreign investors. TRIPS also creates obstacles to the transfer of knowledge through reverse engineering of products.

These trade and investment policies set the stage for FDI in the manufacturing sector to be the engine of Mexican development. There were also changes in domestic policies in order to align the manufacturing sector with the new, neo-liberal macroeconomic, trade, and investment policies. In a marked split from the past, Mexico's overarching approach to industrial policy took a "horizontal" approach. Rather than targeting a handful of firms and industries as it had done under ISI, the state was to treat all firms and sectors equally without preference or subsidy. In a horizontal fashion, the state liberalized imports along with exports, phased out subsidies and price controls, and privatized many SOEs (Dussel 1999, 2003). More specifically, the Mexican government

• provided information services for production and marketing of exports to the manufacturing sector as a whole,

• eliminated price controls,

• shifted the emphasis of Mexico's development banks toward lending at market rates rather than in preferential terms chosen by the state,

• promoted the establishment of industrial clustering to generate and capture knowledge spill-overs from FDI,

• provided regional consulting services and specialized courses for 100 percent Mexican-owned small and medium-size enterprises,

• through its development banks, offered loans and guarantees for demonstration projects and processes to facilitated linkages and spill-overs, and

• tightened government policies toward organized labor by limiting contract negotiations solely to government-friendly unions.

Since the manufacturing sector was to become a low-wage export platform, Mexico also formed a new Program for Industrial and Foreign Trade Policy (PROPICE in Spanish) to manage the integration process and to provide buffers to the vulnerabilities that accompany it. PROPICE reiterated that the goals are to increase productivity, competitiveness, and employment in manufacturing, as well as improve income distribution. PROPICE also stressed the importance of building corresponding supplier networks through small and medium-size enterprises (SMEs), and recognized the need to prepare for competitive shocks from other, lower-wage producers such as China.

Mexico's swing toward economic integration was all but completed under the two presidential administrations serving after Salinas. However, it has not been a smooth arc. Positive macroeconomic results from 1987 to 1993 had led most politicians and analysts to believe that Mexico was well on the way to recovery and industrial restructuring.

In January 1994, Mexico left the G-77 organization of developing countries and joined the "club of rich countries," the OECD. However, in December Mexico spun into another peso crisis that shocked politicians, analysts, and pundits. The shock was followed by investment panic (Edwards 1998). In hindsight, Mexico's stabilization strategies from 1987 to 1994 were said to have led to an overvaluation of the exchange rate, a poor macroeconomic situation, and lack of growth (Dornbusch 1994). Simply stated, Mexico had an overvalued peso and when reserves dried up, investors fled and the peso sank (Dornbusch 1994; Sachs 1995; Pastor 1998).

Like de la Madrid, President Ernesto Zedillo (1994–2000) sought the international community's help to finance domestic reforms. The United States, the IMF, and the Paris Club again provided international financial assistance, this time to the tune of $53 billion. On the domestic front, real spending was reduced, fiscal policy was tightened, monetary growth was limited, and the exchange rate was floated to allow for further depreciations if necessary. This package halted peso devaluation, at least for the time being, and restored the confidence of the all-important foreign investors (Pastor 1998). President Vicente Fox continued the economic policies of his last two predecessors, signing numerous bilateral trade pacts with other nations, pushing for a hemisphere-wide trade and investment agreement, and hosting the fifth ministerial meeting of the WTO in 2003.

Economic Performance

How well did the neo-liberal policies perform? On the one hand, the strategy met the objectives defined by Salinas. Foreign investment and exports surged, and inflation has seemingly been tamed. Yet the underlying long-term capacities of Mexico's domestic firms to learn, to innovate, and to produce for both global and domestic markets deteriorated. Despite surges in foreign investment and exports, Mexico's economy failed to generate significant growth during the liberalization period.

From a macroeconomic point of view, the rate of GDP growth is the single most telling—and troublesome—indicator of how the FDI-led integration strategy affected sustainable industrial development in Mexico. Between 1994 and 2002, GDP grow at an average rate of only 2.7 percent per year. Indeed, GDP growth in the 1990s was less than half the 6.7 percent average growth rate under the ISI policies of the 1970s. Even in the tumultuous 1980s, GDP grew an average of 3.7 percent per year (table 2.1).

Potentially, the troubling low growth performance could be to Mexico's long-term benefit if long term productive capacities were being put in place. On the other hand, if the integration strategy both reduced growth and undermined local capacities for production and innovation, then Mexico's long term economic health is in question. In this section of the chapter we discuss Mexico's success in attracting FDI and increasing exports, and then we examine what happened to three key indicators of productive capacity: (1) total investment (gross fixed capital formation), (2) the capture of knowledge spill-overs from foreign firms, especially through backward linkages, and (3) growth in local capacities for innovation.

Operation Successful? FDI Inflows

Mexico's ability to attract FDI in the post-NAFTA period has been impressive. Among the developing countries that received the lion's share of FDI in the 1990s, only China and Brazil received more than Mexico. Indeed, China, Brazil and Mexico together received 55 percent of all FDI inflows to developing countries between 1990 and 2001 (UNCTAD 2002).

Over the period 1985–2005, foreign investment grew by a factor of 7 and has averaged more than $12 billion per year since NAFTA (table 2.1). Twenty-eight percent was in the form of mergers and acquisitions, while 72 percent was greenfield investment. The spatial distribution of FDI in Mexico has been very uneven, tending to concentrate in urban areas clustered around Mexico City or the US-Mexico border. Since 1994, Mexico City has received 60 percent of the total. Together and in descending order, Nuevo Leon, Baja California, Chihuahua, Tamaulipas, and Jalisco receive nearly 30 percent of all FDI in Mexico.[2]

Manufacturing and financial services accounted for nearly 75 percent of all FDI inflows into Mexico between 1994 and 2002. Agriculture, mining, and construction each received less than 1 percent. Despite their preferential access to imports, maquiladoras received only 32 percent of the total between 1994 and 2002. The vast majority (72 percent) of maquiladora investment flowed to the automotive, electronics, and apparel assembly sectors.

Within the manufacturing sector, nearly half of FDI went to machinery and equipment industries, which include automobiles, electron-

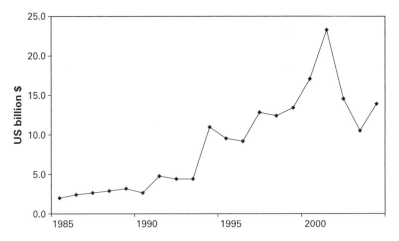

Figure 2.1
Foreign direct investment in Mexico, 1985–2005. Source: World Bank 2006.

ics, apparel, and textiles. Food and beverages and chemicals were the second- and third-largest recipients, with 18 and 13 percent of the total, respectively.

The United States is by far the largest source of FDI in Mexico, accounting for 67 percent of all inflows since 1994. Financial services received the largest amount of US FDI into Mexico. However, these percentages reflect an outlier year. In 2001, Citigroup purchased Banamex for $12.5 billion, accounting for more than half of all Mexico's FDI inflows in that year (UNCTAD 2000). Excluding that year, manufacturing is the leading sector that receives FDI from the United States. In the manufacturing sector, the biggest recipients of US FDI are automobiles, electronics, and clothing.

Reflecting the huge increase in FDI inflows, exports increased by more than a factor of 3 between 1994 and 2002, rising from $50 billion to $160 billion. Eighty-eight percent of Mexico's exports were from the manufacturing sector, distinguishing Mexico from many other Latin American countries, such as Chile, which remain heavily dependent on minerals and other primary products. Metallic products, equipment, and machinery—which include autos and electronics—accounted for 72 percent of manufacturing exports and about 64 percent of all exports. Manufactured exports grew at the rapid clip of 13.8 percent a year on average between 1994 and 2002.

Despite large increases in FDI and manufactured exports, the long-term stability of Mexico's integration strategy is far from assured. There are two overarching sources of instability. First, imports grew even faster than exports between 1994 and 2002, generating a large and persistent current-account deficit. The manufacturing sector ran an average $11.4 billion deficit during the period 1994 to 2002, accounting for approximately 80 percent of the deficit. Both foreign and domestic manufacturing firms producing for export rely overwhelmingly on imported, rather than locally sourced inputs. Indeed, the share of locally sourced inputs in maquila plants dropped from 4.7 percent to 3.7 percent between 1990 and 2002 (table 2.2).

Undoubtedly, a series of complex and interrelated factors drive the reliance on imports in Mexican manufacturing. One, however, is the fact that Mexico's exchange rate is overvalued, itself the result of a high

Table 2.2
Locally sourced inputs in *maquila* manufacturing plants (value added as percentage of total). Source: INEGI 2003.

	1990	2002
Food and beverage	38.6	42.9
Apparel	0.9	8.0
Footwear and leather	6.9	2.7
Furniture and wood products	3.1	14.6
Chemical products	10.2	5.5
Transportation equipment	0.8	3.3
Machinery	4.6	2.6
Electronic assembly	1.2	3.0
Materials and electronic accessories	0.01	2.2
Sporting goods	1.1	2.2
Other	4.1	2.2
Average	4.7	3.7

interest, anti-inflation policy (Nadal 2003). While inflation was largely brought under control, the cost was a ballooning current-account deficit (Dussel 1999). Second, Mexico's low-wage competitiveness has begun to slide. According to the World Competitiveness Yearbook, Mexico fell from 34th place in 1998 to 41st in 2002. In another index operated by the World Economic Forum, Mexico slipped from 42nd to 48th place from 2000 to 2004. Yet another ranking, the Microeconomic Competitiveness index, put Mexico at number 42 in 1998 and 55 in 2002. Falling competitiveness has been attributed to three factors: the economic slowdown in the United States, the relative strength of the peso, and other factors such as the emergence of China's entry in the WTO (CSIS 2003; Gerber and Carrillo 2003).

In terms of the performance of its "foreign sector," the integration strategy, in short, has an Achilles' Heel. While it has achieved some of its central goals, including controlling inflation and increasing exports and foreign investment, persistent current-account deficits and overvalued exchange rates suggest that the strategy is not sustainable in the long term. The Mexican economist Enrique Dussel sums up the situation: "The export sector's overall inability to generate linkages with the rest of the economy—in terms of employment, learning processes, and technological innovation, among many other aspects—creates unsustainable macroeconomic conditions in the medium and long term. As soon as the economy (particularly through manufacturing) grows in terms of GDP and exports, it requires larger quantities of imports for capital accumulation." (2003, p. 270)

The reliance by MNCs on imported inputs choked off the potential benefits of FDI for sustainable development.

Crowding Out Domestic Investment

The promise of a successful FDI-dependent strategy is that it will add more than its own value to total investment by "crowding in" domestic investment. Through a variety of channels, the hope is that FDI will stimulate domestic investment in foreign firms or domestic firms that compete with or supply them. In Mexico however, FDI had the opposite effect: FDI "crowded out" domestic investment.

Despite the large FDI inflows, total annual investment as a percent of GDP was the same between 1994 and 2002 as in the 1980s and down a bit from the 1970s. In both periods, it averaged 19.4 percent a year. UNCTAD argues that to "catch up" with OECD countries, developing countries need total investment to reach 25 percent of GDP. But the share of FDI in total investment more than doubled, rising from 5.4 percent between 1981 and 1993 to 12.6 percent between 1994 and 2002. The share of domestic investment plummeted from 14 percent to 6.8 percent.

The manufacturing sector has apparently been hard hit by the contraction in domestic investment. From 1970 to 1982, investment in manufacturing averaged about 10 percent of GDP and accounted for nearly half of total investment. In the 1980s, investment in manufacturing dropped off to just over 5 percent of GDP, accounting for just over 25 percent of total investment. While more recent data is not available, data from 1988 to 1994 show the persistence of a contractionary trend: despite the fact that nearly half of FDI inflows went into the manufacturing sector, total investment remained under 6 percent of GDP (Moreno-Brid 1999).

Rather than stimulate new investment, FDI and the liberalization strategy overall, apparently "crowded out" domestic investment.

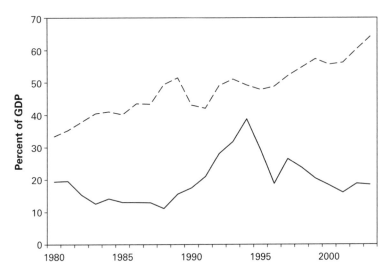

Figure 2.2
Domestic credit to the private sector: Mexico (—) vs. other middle-income countries (- -). Source: World Development Indicators 2004.

Crowding out was not due to excessive borrowing by trans-national corporations (TNCs) in domestic capital markets. Generally, TNC affiliates and domestic companies producing primarily for export have access to foreign sources of finance. According to many economists, the overarching cause was the anti-inflationary macroeconomic policy package, which generated high interest rates and an overvalued exchange rate (Nadal 2003). A key element of the package, which aimed to suppress aggregate demand, was contractionary monetary policy. A high prime rate pushed up commercial bank interest rates, which averaged 22 percent between 1994 and 2002. The resulting credit crunch is illustrated in figure 2.2.

High interest rates choke off domestic investment directly, by raising the cost of capital, and indirectly, by leading to an overvalued exchange rate, generated by inflows of foreign capital attracted by high interest rates. Indeed, a central objective of the liberalization strategy is the attraction of foreign portfolio capital inflows to finance balance of payments gaps. Moreover, the government has made the exchange rate the anchor of its domestic price system and undertakes interventions to raise the value of the peso, even though it is supposed to float (Nadal 2003). An overvalued

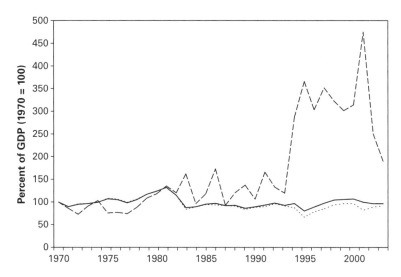

Figure 2.3
"Crowding in" domestic investment? – –: foreign. - -: domestic. —: total. Source:
World Development Indicators 2006.

exchange rate makes imports, including of intermediate products, cheap
relative to domestic production. Domestic producers get crowded out.

Besides high interest rates and overvalued exchange rates, Mexican
policies have constricted aggregate demand by constraining wage
growth through "economic solidarity pacts." Real wages in manufacturing outside of the maquiladoras have decreased by 12 percent since
1994 (Arroyo 2003; INEGI 2003; Salas 2003). While the pacts helped
to control inflation, they also drove incomes down. Domestic demand
for manufactured goods plummeted after 1994 and growth remained
sluggish throughout the decade. Between 1994 and 2001, domestic sales
of manufactured goods rose by only 22 percent, while export sales rose
by 212 percent (in 1995 dollars).

Missing Spill-Overs

As we discussed in chapter 1, the central "promise" of FDI is that it
delivers "knowledge spill-overs"—increases in technology and skills—to
local firms and workers. By generating improvements in technology, efficiency, and productivity, FDI stimulates economic growth (Lim 2001).
Did FDI generate spill-overs in Mexican manufacturing?

In a case study of the automotive sector in Mexico, Moran (1998, pp. 53–56) found that the integration of Mexican producers into the global sourcing and marketing strategies of multinational car companies in the 1980s generated a host of spill-overs to local firms and local communities. Spurred by the government's export performance requirements, the big automakers—led by General Motors—invested heavily between 1979 and 1981. Productive capacity grew rapidly—the production of engines alone grew to more than a million units per year. Employment in the auto industry swelled and wages and benefits were among the highest in the country.

The decision to produce for export rather for the Mexican market, as they had done under domestic content requirements, led the automakers to transfer best production technology and to introduce industry best practices, such as zero-defects procedures and production audits. According to evidence cited by Moran, the backward linkages were extensive: within 5 years, there were 310 domestic producers of parts and accessories, of which nearly one-third had annual sales of more than $1 million. There were also spill-overs of export marketing skills: only four of the ten largest auto parts exporting firms in 1987 had foreign ownership.

Other studies, however, are less optimistic. In a large statistical study covering 52 Mexican industries, Romo Murillo (2003) examines four types of spill-over mechanisms: backward linkages, collaboration effects (e.g. joint R&D), demonstration effects, and training effects. He found that foreign presence is positively related to demonstration and to training effects, but negatively correlated with demonstration effects. Most important, he found no evidence that foreign presence is linked to technology spill-overs (Romo Murillo 2002).

The reason that knowledge spill-overs from FDI are hard to find in Mexico may be due to the fact that, apart from the auto industry, Mexican suppliers remain largely out of the subcontracting loop. In only a few industries—food and beverage and furniture and wood products—do locally sourced inputs account for a significant share of total inputs (table 2.2). However, these industries are relatively small. Overall, the share of locally sourced inputs in manufacturing contracted between 1990 and 2002.

Besides squeezing out possibilities to gain efficiency spill-overs, the lack of local sourcing means that the export-oriented strategy generates a persistent current-account deficit. Between 1994 and 2002, the excess of imports over exports averaged $13.5 billion per year (INEGI 2003).

Knowledge and Innovation

The growth of endogenous productive capacities, especially the capacity for innovation, requires investment in expanding and utilizing knowledge. Knowledge is required to absorb new technologies, be globally competitive in cutting-edge industries, and to design and market new products and services, in domestic or global markets.

Especially for firms in "latecomer" countries like Mexico, investment in knowledge is the "make it or break it" variable which determines whether firms can compete in mature industries, which earn thin and declining margins. "Even if a firm starts small, " Amsden and Chu (2003, p. 3) conclude in a study of Taiwan's successful high-tech industry, "it must ramp up very quickly to achieve a high output level, a process that requires building assets related to project execution, production engineering, and a form of R&D that straddles or falls somewhere in between applied research and exploratory development." To nurture the capacity for innovation, investment is needed by both the public and private sectors in assets related to project execution, engineering and R&D. Indeed, Amsden and Chu argue that the most important factors in Taiwan's success in the high-tech industry were government subsidies for R&D channeled to nationally owned firms. The government undertook R&D in its own laboratories, initiated joint research projects with the private sector, and subsidized private R&D (Amsden 2003, p. 12).

How does Mexico perform against the yardstick of "capacity for innovation"? A thorough analysis of Mexico's R&D policies or other indicators are beyond the scope of this chapter, but a series of snapshots comparing Mexico and South Korea are telling. In 1981, 648 scientific journal articles were published in Mexico, compared to 168 in South Korea. Between 1995 and 2000, R&D as a percentage of GDP averaged 0.36 percent in Mexico. For manufacturing, R&D as a percentage of manufacturing GDP in Mexico is even lower, at 0.22 percent (Dussel 2004). In South Korea, it was nearly 10 times greater, averaging 2.6 percent. In the same period, scientists and engineers per million people averaged 225 in Mexico, versus 2,152 in South Korea, and R&D

Table 2.3
Capacity for innovation in Mexico and South Korea (average, 1995–2000).
Source: World Bank 2005.

	Mexico	South Korea
Patent applications—resident share of total	4.91%	51.00%
R&D expenditure as % of GDP	0.36%	2.60%
Scientists and engineers per million people	225	2,152
Science and technology journal articles	2,024	5,219
R&D technicians per million people	172	576

technicians per million people averaged 172 in Mexico, versus 576 in South Korea (table 2.3).

Another indicator of innovation capacity is the number of patent applications by residents. In 2000, Mexican residents applied for 451 patents, an increase of some 16 percent over 1996. Non-residents, on the other hand, applied for 66,465 patents in 2000, an increase of nearly 120 percent over 1996. Indeed, the resident share of total patent applications fell by a half, dropping to only 0.67 percent in 2000. In South Korea, in contrast, the resident share of total applications averaged 51 percent between 1995 and 2000 (World Bank 2003). In Taiwan, the resident share over the same period was 75 percent (Amsden 2003).

Employment

Between 1994 and 2002, a total of 637,000 new jobs were created in the manufacturing sector, or about 82,500 each year. However, according to estimates provided in Mexico's national accounts, roughly 730,000 "new entrants" were added to the economically active workforce each year, for a total of 6.5 million *new* workers between 1994 and 2002. The manufacturing sector, in short, provided jobs for less than 12 percent of the people newly seeking employment each year. Moreover, job growth in the manufacturing sector has been declining since 1997.

Not surprisingly, given the "crowding out" of domestic investment, nearly all the new jobs—nearly 96 percent—were in the maquila sector. Despite accounting for about two-thirds of total investment, the major exporting firms and maquiladoras accounted for only 5.8 percent of total employment in Mexico in 2002 (Dussel 2003). The foreign sector, in other words, creates a small number of manufacturing jobs in enclaves across Mexico. Moreover, jobs in the foreign sector are vulnerable to competition from lower-wage export platforms, especially China, and to changes in global markets, especially slowdown in the US economy.

There is evidence that the jobs created since 1994 are of poor quality. According to national employment surveys published by INEGI, 55.3 percent of new jobs do not provide benefits. Indeed, 49.5 percent of the employed Mexican workforce is without benefits (INEGI 2003; referred to in Arroyo 2003). The minimum wage in Mexico declined by more than 70 percent since 1982 and 7 percent since 1994 (INEGI 2003).

Despite an 18 percent increase in productivity, wages in Mexican manufacturing overall have declined by 13 percent since 1994. Manufacturing wages gained ground between 1987 and 1994 but collapsed as a result of the 1995 peso crisis. In real terms, wages in 2002 were 24 percent lower than in 1982 (Salas and Zepeda 2003).

In keeping with the policy of low-cost/low-wage manufacturing, wages in maquilas are lower than in the manufacturing sector as a whole. Real wages in maquiladoras averaged less than 80 percent of wages in non-maquiladora manufacturing between 1987 and 1994 (Alcalde 2000). Maquila wages have increased relative to non-maquila wages since 1994 but were still 14 percent below the non-maquiladora manufacturing wage in 2002 (INEGI 2003). One study found that wage gains in Mexico have been the largest in those firms most exposed to international trade and investment (Hanson 2003). In other words, the one area where wages increased in the 1990s was in the foreign enclave.

Environmental Performance

Many environmental trends are worsening in Mexico. Between 1985 and 1999, the economic costs of environmental degradation—including rural soil erosion, municipal solid waste, and urban air pollution—amounted to 10 percent of annual GDP, or $36 billion per year (INEGI 2000; Gallagher 2003; Gallagher 2004). These costs dwarf overall economic growth, which amounted to only 2.6 percent on an annual basis (INEGI 2000). Indeed, these damage cost figures were cited by the World Bank as part of the rationale for a new environmental loan to Mexico in 2002.

In this section we examine the environmental performance of the Mexican manufacturing sector under the FDI-led strategy. Although overall trends are worsening, there is some evidence of environmental improvement through compositional effects and through technology transfer from foreign firms. We also examine the role of the Mexican government in promoting better environmental performance through increased compliance with existing regulations. Finally, we consider the extent to which the largest US TNCs in Mexico are embracing voluntary initiatives to improve their environmental performance under the mantle of corporate social responsibility.

Exports and the Environment

Two recent in-depth studies evaluate the environmental impacts of export-led manufacturing growth in Mexico. Both come to similar conclusions: overall levels of industrial pollution, particularly criteria air pollution, water pollution, and toxics, have increased faster than population growth and faster than the GDP of the economy as a whole in Mexico since the 1980s. Both studies find that environmental degradation was fueled by large increases in manufacturing growth and exports. In other words, the overall "scale" of economic activity in the manufacturing sector corresponded with a growing amount of pollution. However, both studies found that overall levels of pollution occurred somewhat slower than overall growth in manufacturing output and overall growth of exports. The relative improvements were due to "composition effects," small shifts away from pollution-intensive manufacturing (Gallagher 2002; Schatan 2002; Gallagher 2004).

Under the integration strategy, Mexico consolidated its comparative advantage in labor-intensive assembly work and sold off state-patronized industries such as steel, cement, and pulp and paper. On the whole, labor-intensive industries are less pollution-intensive than their heavily capital-intensive counterparts in the manufacturing sector. This explains why these studies have found compositional shifts away from pollution-intensive industry. However, both studies point out that such "compositional" changes toward relatively less pollution intensive industry have been far outweighed by overall scale effects of rapid industrial growth. One of the studies predicts that for every 1 percent increase in manufacturing output there was a corresponding 0.5 percent increase in pollution; the other study examines criteria air pollution only and predicts a corresponding pollution growth rate of 0.7 percent (Schatan 2002; Gallagher 2005).

Two other studies, by the OECD and CEPAL, examined the foreign-dominated maquiladoras in particular. These studies note that the "on-site" pollution of maquila assembly plants is relatively less pollution-intensive than their heavy industry counterparts. However, maquila growth attracts workers and stimulates rapid migration. The population influx far exceeds the infrastructure capacity of host communities and has led to inadequate management of sewage and waste, insufficient supplies of water, and negative consequences for air quality (OECD 1995; Stromberg 2002).

In addition, air pollution, once seen as problematic only in Mexico City, is becoming problematic in all major industrial cities. Guadalajara now exceeds air pollution norms for 40 percent of the year, Monterrey for 25 percent, Ciudad Juarez for 7 percent, Mexicali for 30 percent and Tijuana for 4 percent. With the exception of Guadalajara, these trends have all worsened since 1993 (Stromberg 2002).

While environmental trends are worsening, there is little evidence that Mexico is a "pollution haven." During the NAFTA debates, there was widespread concern that pollution-intensive US firms would relocate to the border to evade tougher US laws. One study found that California-based furniture makers moved to Mexico to avoid installing air pollution fixtures, and Mexico reportedly made statements attempting to lure US firms by making low regulatory compliance costs part of their sales pitch (Mayer 1998).

On the whole, however, there has not been a "giant sucking sound" of dirty industry flocking to Mexico. Between 1988 and 1998, the share of "dirty industry" in total manufacturing production fell by 3.6 percent in Mexico and by 2.3 percent in the United States. Employment in dirty industries in the United States remained about the same and declined by 2 percent in Mexico.

A number of empirical studies have similarly concluded that Mexico is not a pollution haven. In a cross-industry comparison of data in one year, 1987, Grossman and Krueger (1993) tested whether pollution abatement costs in US industries affected imports from Mexico, as one would expect if Mexico was a pollution haven relative to the United States. They found the impact of cross-industry differences in pollution abatement costs on US imports from Mexico to be positive, but small and statistically insignificant. Indeed, traditional economic determinants of trade and investment, such as factor prices and tariffs, were found to be far more significant.

A more recent study examined whether pollution abatement costs affected patterns of US foreign investment into Mexico and three other countries (Eskeland and Harrison 1997). The authors found a statistically insignificant, though positive, relationship between pollution abatement costs and levels of FDI. In a time-series study, Kahn (2001) examined the pollution intensity of US trade with Mexico and other countries. Using US Toxic Release Inventory data for 1972, 1982, and 1992, he found the pollution content of US imports from Mexico slightly declined during the period.

A leading explanation for environmental improvements is the behavior of foreign firms. Three studies conclude that foreign presence in the Mexican steel industry led to better environmental performance. Gentry and Fernandez (1998) found that Dutch steel firms and the Mexican government brokered an agreement whereby the Mexican government agreed to share some of the environmental liabilities of the sector. Later, the foreign firms began investing in environmental improvements. A broader study of the Mexican steel sector found that foreign firms, or firms that serve foreign markets, were more apt to comply with environmental regulations in the steel sector (Mercado 2000).

A third study which examined criteria air pollution in Mexican steel, found that the Mexican sector is "cleaner" per unit of output that its US counterpart. This is partly due to the fact the new investment (both foreign and domestic) came in the form of more environmentally benign mini-mill technology rather than more traditional and dirtier blast furnaces. Based on this analysis, the author hypothesized that when pollution is in large part a function of core technologies, new investment can bring overall reductions in pollution-intensity. However, when pollution is a function of end-of-pipe technologies, new investment will not necessarily correspond with reductions in pollution intensity unless such technology is required and enforced by government (Gallagher 2004).

Standards and Compliance: The Role of the Mexican Government

Dirty industries did not relocate to Mexico en masse as a result of NAFTA. On the other hand, Mexico offered a generally laxer climate of environmental regulation for all industries than many US states. In some cases, environmental standards were lower or non-existent.

In other cases, however, Mexican standards were—and are— relatively high, the result of significant evolution of environmental awareness during the 1990s. The problem is lack of enforcement, stemming in part from Mexico's macroeconomic and fiscal crises. While they may not have been drawn to Mexico because of lower environmental standards, foreign firms would have had the opportunity to perform poorly once they got there. No doubt some did.

To assess the environmental performance of the FDI-led integration strategy in general, the issue of compliance by firms—domestic and foreign—with environmental regulation is paramount. What are the determinants of regulatory compliance in Mexico?

Two World Bank studies concluded that the important determinants of compliance by domestic and foreign firms with environmental regulations in Mexico are (1) government pressure, including inspections, (2) local community pressure, and (3) whether or not the firm has an environmental management system (Dasgupta et al. 1997). Interestingly, one of the studies found no correlation between compliance and foreign origin (Dasgupta et al. 2000). Foreign firms, in other words, were no more likely to comply with regulation than domestic firms.

When foreign firms *are* in compliance, one study has shown that regulation and inspections are important determinants. A survey of 44 US manufacturing firms in Mexico showed that environmental improvements such as investing in water treatment facilities were motivated by regulation and enforcement by the Mexican authorities (Gentry 1998). A very recent study of 222 manufacturing firms in Mexico also found regulatory pressure to be the most significant driver of environmental performance. However, that same study also found firms exporting to the United States and Canada were more apt to be responsive to environmental concerns than non-exporting firms (Wisner and Epstein 2003)

Despite its efficacy, there are signs that Mexico's commitment to regulatory pressure may be falling by the wayside. Although spending on environmental protection grew impressively between 1988 and 1993, it tapered off by 45 percent between 1994 and 1999 (Stromberg 2002). Although such spending on the environment has grown considerably compared to earlier levels, it remains the lowest of all OECD countries. In relation to GDP, the average OECD country spends 3 times as much as Mexico on the environment. Per capita, the average OECD country spends 6 times as much as Mexico (OECD 1998).

Environmental inspection patterns mirror the trend in environmental spending. Although inspections got off to an impressive start in 1992, only 6 percent of establishments were inspected at the highest point. Total inspections decreased by 45 percent after 1993, and inspections in the maquila sector decreased by 37 percent.

The implementation of environmental management systems (EMS) has been found to correlate with firm-level environmental compliance. Although they are becoming more popular, the number of Mexican firms with EMS still remains very small. According to industry sources, 266 Mexican firms were certified to ISO 14,001, the international EMS standard, as of 2002—only 0.1 percent of all firms. Countries such as Brazil, Korea, Taiwan, and China have between 3 and 5 times the number of ISO certifications as Mexico (ISOWorld 2003).

Corporate Social Responsibility?

Prodded by pressures from environmental and community groups, as well as the threat of regulation, a number of companies, both foreign and Mexican, have taken voluntary initiatives to improves their environmental and social performance. Under the mantle of "corporate social responsibility," they have generated codes of conduct, implemented environmental management systems, consulted with advocacy groups, and/or produced "sustainability reports" disclosing information about company environmental performance

Voluntary initiatives have helped to improve company communication with the public. How effective they are generally in improving environmental and social performance remains a subject of study (Leighton et al. 2002). In Mexico, a handful of studies have shown that foreign firms have transferred environmentally friendly technology and management methods to Mexico.

One study described the way that affiliates of US chemical firms teamed up with the Mexican chemical industry to incorporate US "responsible care" environmental policies into operations of the Mexican chemical industry (Garcia-Johnson 2000). Another study on the chemical fibers industry found that although environmental regulations and inspections were the key driver for environmental compliance in that industry, foreign participation in the industry was correlated with environmental improvements as well (Dominquez-Villalabos 2000).

There are signs that some portions of the Mexican business community are beginning to take the environment more seriously. In 1992, Mexico's National Council of Ecological Industrialists (CONIECO) was created as an organization of manufacturers and resellers of products that can help clean the environment. The Latin American chapter of the World Business Council for Sustainable Development was established in Mexico City in 1993. In 1994, the Center for Private Sector Studies for Sustainable Development (CESPEDES) was formed (Barkin 1999). And in 2002, the Mexican cement giant CEMEX received the 2002 World Environment Center's Gold Medal for International Corporate Environmental Achievement.

To what extent have foreign companies voluntarily adopted good environmental management as part of "standard operating procedure" in Mexico? In the absence of studies, we examined the "Broad Market Social Index" (BMSI) ratings of the largest publicly traded US firms operating in Mexico.

Produced by KLD Research and Analytics Inc, a Boston-based firm, the BMSI—a proxy for "corporate social responsibility"—is created by screening companies according to four criteria.[3] Of the 26 largest (by total sales) US firms operating in Mexico in 1999, 14 met the BMSI criteria, including IBM, PepsiCo, Motorola, Hewlett-Packard, and Procter & Gamble. Among the 12 that did not meet the criteria were the three largest US companies in Mexico: General Motors, Ford, and Wal-Mart.

This chapter has provided a broad overview of the profound transformation that has occurred in Mexico's economy and demonstrated how this transformation has resulted in the emergence of an "enclave economy." The question is whether the IT sector in Guadalajara is the exception or the rule.

3

Globally Networked, Environmentally Challenged: A Profile of the IT Industry

More than on any other industry, Mexico pinned its post-NAFTA hopes for economic development on the local growth of a vibrant, export-oriented information technology industry. Building on geographic proximity to the United States, as well as manufacturing and supply capacities developed during the ISI period, the idea was that an influx of foreign direct investment by American multinational corporations would make Mexico the IT manufacturing hub of the Americas. From their Mexican base, a nexus of MNCs and local firms would manufacture final products for export to voracious consumers in the north, and serve as a design and product introduction gateway for emerging markets in the south.

Mexico was not alone in setting its development sights on the IT industry. In the 1990s, high-tech products were the fastest-growing sector of manufactured exports from developing countries as a whole. By 1998, developing countries' exports of electronics outstripped exports of textiles, clothing and footwear by more than 50 percent (UNIDO 2002). Beyond generating jobs and export earnings, the knowledge-intensive IT sector was seen as an agent of industrial transformation and sustainable development—a driver of more productive, globally competitive and environmentally cleaner national economic growth.

A central goal of this book is to untangle the reasons for the failure of the IT promise to be fulfilled in Mexico in the 1990s, and to glean policy lessons that could help Mexico and other developing countries to garner greater economic, social and environmental benefits in the future. With this chapter, we begin our case study by profiling the global IT industry in three key dimensions: industrial organization, global geography of IT manufacturing, and environmental challenges. For developing-country governments, an understanding of how the industry operates, where it locates, and the environmental hazards it poses is crucial in designing policies which minimize harm and maximize development benefits of FDI by IT firms.

Global Production Network

Spanning from televisions to computers to cell phones and video games, the global electronics industry is second only to automobiles and transportation equipment as the most traded sector in the world economy.[1] Mexico and many other nations divide the huge electronics industry into three broad subcategories: audiovisuals (televisions and so forth), the domestic appliance sector (radios and appliances), and information technology.[2]

Information technology is the fastest-growing part of the electronics industry. As an industry category of its own, IT broadly refers to computers and peripherals, servers, monitors, printers, software development and design, and, to some extent, electronic telecommunications (e.g. cell phones). With the design of new applications, communications and information technologies are increasingly collapsing into each other, causing industry boundaries to blur.

Since the late 1980s, the electronics industry as a whole has been characterized by an increasing degree of outsourcing, that is, the contracting out of key productive functions to external firms. Starting in the mid 1990s, the industry has also by a high degree of offshoring (outsourcing across national borders, especially in developing countries). Taken together, outsourcing and offshoring is what is meant by "globalization."

In any industry, there are three main parts to a value chain (Dedrick 2002a): product innovation (R&D, design, market research and new product introduction), operations (process engineering, manufacturing, logistics, finance and human resources), and customer relations (marketing, sales, advertising, distribution, customer service, technical support). In the traditional vertical model of industrial organization, all three functions are undertaken within a single firm. From the 1960s to the 1980s, IT giants such as IBM and Hewlett-Packard were, for the most part, structured vertically. Offshore assembly and components manufacturing started in the 1960s and grew substantially in the 1970s, especially in Southeast Asia. However, with the exception of Taiwan, manufacturing functions were not outsourced but were undertaken within MNC affiliates. (See chapter 4.)

Starting in the late 1980s, large IT firms increasingly began to shed their manufacturing functions and outsource them to external firms. Today, the IT industry is organized not in vertically integrated firms but in a "global production network" (GPN): independent firms undertake productive functions linked together through external transactions.

The firms in a GPN, also referred to as a "global value chain," range from very large, highly concentrated TNCs to small and medium-size, even mom-and-pop, enterprises. Firms enter and participate in a GPN in different parts of the value chain (figure 3.1). In broad terms, a GPN has

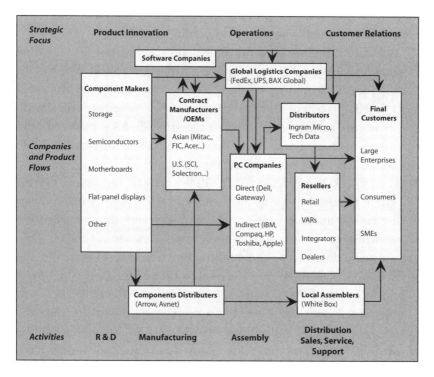

Figure 3.1
The PC value chain. Source: Dedrick 2002a.

three types of firms: the global flagship or GPN leader, a small number of large contract manufacturers, and a large number of small and medium-size suppliers.

Global Flagships: Brand-Name Companies

At the heart of the global production network is the global "flagship"—a large company (such as Hewlett-Packard, IBM, Toshiba, or Dell) that provides strategic and organizational leadership and a brand name to the GPN.[3] While manufacturing and other productive functions are outsourced, flagships retain in-house activities in which they have a particular strategic advantage and which earn high returns (Ernst 2001).

Typically, the flagships have kept two key functions inside the company: strategic marketing and product innovation, including R&D and design.[4] The flagship brand and its global sales capacities are essential to the marketing function. Along with the flagship's role in coordinating transaction and knowledge exchange throughout the GPN, these two functions are the major sources of leadership and profits.

A handful of global flagship firms hold the vast portion of global IT market share, making the industry a highly concentrated oligopoly. Economists call an industry an oligopoly when the four largest firms hold more than 25 percent of overall sales (Blair 1972). "Highly concentrated" oligopolies are said to be those where the top eight firms control 70–85 percent of the market and the top four control 50–65 percent. By 2000, the global IT industry was more concentrated than the world oil industry at its cartel peak. Examples include hard-disk drives (five firms account for 85 percent of sales, head assembly (ten manufacturers control 93 percent of the market, and the largest six have 78 percent), dynamic random-access memories (the top four firms control 66 percent of the market), and personal computers (in 2004 the top four firms had 44 percent of the worldwide market—up from 27 percent in 1996—and the top four firms had 63.7 percent of the US market) (Ernst 2000; Markoff 2003).

Combined with the huge investments required to get into the industry, the strategic capabilities required of flagships act as barriers to entry, protecting and enhancing industry concentration. Led by IBM, US firms consolidated their position as oligopolistic industry leaders in the 1960s and the 1970s. Only with the help of government policies, including high tariffs in Japan and anti-trust lawsuits against IBM in the United States, were Japanese firms able to "break in" in the 1970s and the 1980s (Chandler 2001).

According to the industrial organization guru Alfred Chandler Jr. (2001), knowledge is the foundation of the commercial strength of the global flagship companies. Their capacities for innovation, coordination, and marketing derive from the continuous evolution of knowledge in three areas:

Technical research capabilities learned from applying basic scientific and engineering knowledge to create basic products and processes

Functional development, production, and marketing capabilities. (Development is the knowledge required to transform basic research innovations into commercially viable products. Production capabilities are the capabilities required to build and organize large scale production facilities that can produce the new products. Marketing capabilities are the skills required to distribute the new products to the right markets.)

Managerial ability to integrate research, design, development, production and marketing across diverse geographical locations.

IBM pioneered a prototype of the flagship model in the 1950s and the 1960s, when it established an integrated, transatlantic R&D and production network spanning Europe and the United States. "Product development and manufacturing responsibilities were assigned to individual laboratories and production facilities; each development laboratory specialized in a particular technology and carried the development responsibility for a product or technology for the entire company. Each IBM plant, including the US facilities, was given a mandate to produce specific products both for the international and the local market." (Ernst 2000)

Contract Manufacturers
For IBM and the other leading electronics companies that followed suit over the next 20 years, internationalization meant offshore production by foreign affiliates. In the late 1980s, however, global flagships began outsourcing standardized, low-value-generating manufacturing operations to independent firms called "contract manufacturers" (CMs). Vertical integration gave way to modular production and vertical specialization.

Table 3.1
Data on top five contract manufacturers, 1995–2002. Source: Sturgeon 2002.

Firm	Revenue 1995	2002	Annual growth rate	% revenue from acquisitions
Solectron	1,680	16,470	46	54
Flextronics	389	13,160	80	72
Sanmina/SCI	3,514	12,140	23	112
Celestica	600	11,250	63	53
Jabil Circuit	686	4,870	39	13
Total	6,869	57,890	43	67

Contract manufacturers assemble components to produce the innards of final products, which are then branded by their flagship customers. Known also as "Electronic Manufacturing Services" (EMS), contract manufacturers today play a very substantial role in the GPN and are the fastest-growing sector in the global electronics industry. Solectron, for example, now the world's largest CM, was a company of modest size in 1995, with revenues of about $1.7 billion. After growing by an average 46 percent per year, analysts predicted revenues of nearly $16.5 billion in 2002.

All five major contract manufacturers are North American companies: Solectron (based in Milpitas, California), Flextronics International (incorporated in Singapore but managed from its headquarters in San Jose, California), Jabil Circuit (based in Saint Petersburg), Sanmina/SCI (based in San Jose), and Celestica (based in Toronto). Between 1995 and 2002, the combined revenues of the top five grew from about $6.9 billion to $57.9 billion. According to a 2000 report by Electronic Trend Publications, the top five had captured an estimated 38 percent of global electronics contract manufacturing by 1999 and their share was expected to grow to 65 percent by 2003 (Sturgeon 2002).

According to Luthje (2003), CMs have five distinctive characteristics:

Manufacturing as service   Since CMs do not manufacture their "own" products, worker management and quality control must be oriented toward their customers rather than toward the final consumers. Manufacturing thus becomes organized as a service industry.

Low wages   The high level of standardization means that CMs generate little value and seek low-labor-cost production sites.[5]

Labor flexibility   Because production volumes change constantly and rapidly, CMs maintain highly flexible employment policies.

Quality control based on task uniformity   Tasks are highly standardized and there is no formal work group structure.

Women and minority workers   CM workers tend to be drawn from low-wage parts of the population, especially women and ethnic minorities.

Operating on a margin of 3–5 percent, CMs earn profits from their capacity to manage complex supply chains to a large scale to suit the particular specifications and timelines of a number of brand-name customers. "How do you survive in an industry with a 3 percent margin?" asked a recent electronics industry e-zine profile of Flextronics, the second-largest CM. The answer? By working to a wide client base and "relentlessly wringing efficiencies from its worldwide supply chain" (Koch 2001). Flextronics, for example, uses a global computer system to organize real-time data on supply prices—over four continents—allowing buyers to pinpoint the lowest price for a particular component at any moment. The ability to pool orders for multiple clients and to gather and analyze supply chain information is at the heart of competitive advantage in the CM industry. Indeed, Flextronics and the other big CMs coordinate such a multiplicity and diversity of transactions they could be seen as global supply flagships in their own right (Ernst 2001).

The trend toward global outsourcing to "turnkey" CMs is likely to continue. In a recent Bear Stearns survey, 85 percent of the brand-name electronics firms interviewed said they planned further increases in production outsourcing. As a whole, the brand-name firms said they expected to outsource 73 percent of production; 40 percent of the firms said they planned ultimately to outsource 90–100 percent of final product manufacturing (Sturgeon 2002, p. 461).

Suppliers of Components

The razor-thin margins of the CMs translate into enormous cost pressure on the next tier of the Global Production Network—local suppliers. Ernst and Kim (2001) distinguish between two types of network suppliers.

Large "higher-tier" suppliers, such as the Acer group in Taiwan, play an intermediate role between the global flagship companies and local small and medium-size firm which supply components and other goods and services. These large supply companies play a similar role as the CMs in managing and coordinating global supply chains. Indeed, in some cases, they may act like CMs in providing turnkey services—management of all value chain functions with the exception of R&D and strategic marketing.

"Lower-tier" local suppliers are the vast number of small and medium-size firms that manufacture a large range of inputs into the manufacturing, packaging and transport of final IT products. Local suppliers are often co-located with CMs, though increasingly CMs procure supplies via short-term contracts based on price from suppliers in many parts of the world. Typically, local suppliers do not interact with the global flagships but with the CMs or higher-tier suppliers.

The competitive advantages of lower-tier local suppliers are low production costs and speed and flexibility of delivery. While global flagships and CMs occupy weighty and fairly stable positions in the GPN—despite intense global competition (see below)—the position of local suppliers is precarious. According to Ernst and Lim (2001, p.11), "local suppliers are used as 'price breakers' and 'capacity buffers' and can be dropped at short notice. . . . [They] normally lack proprietary assets; their financial position is weak; and they are highly vulnerable to abrupt changes in markets and technology, and to financial crises."

Drivers of the Global Production Network Model

Global Production Networks, characterized by a high degree of external contracting of key productive functions, represent a decisively new model of industrial organization. While other industries have followed suit, developing their own variant of a GPN, the electronics industry pioneered the model. The emergence and adoption of the model by electronics firms in the 1990s was driven by economic liberalization, innovation in communications and transport technologies, the emergence of standardized components and generic manufacturing capacities in the electronics industry, and the dynamics of cutthroat global competition in the IT industry (Ernst 2001; Sturgeon 2002).

The global shift toward economic liberalization and privatization in the late 1980s and the early 1990s set the stage. As more and more countries adopted open policies toward trade, FDI, and capital flows, all TNCs, not just in the IT sector, faced a more uniform, predictable, and market-oriented global economy. Liberalization changed the choice set for strategic firm decisions, allowing TNCs (indeed forcing them) to consider a wider range of countries as possible manufacturing and marketing sites.

The communications, information, and transport technologies and products generated by the IT sector itself promote global production and marketing chains. On the one hand, they facilitate the global coordination of huge amounts of information and agents. On the other hand, the very large costs of R&D must be amortized over very large, global markets.

The availability of standardized components allowed computer design to shift away from centralized to decentralized architectures. Today, personal computers, microprocessors, memories, and software are organized into product-specific value chains. At the same time, the manufacturing capacities required to produce a wide range of electronics products increasingly became generic, rather than product-specific. Generic manufacturing capacities allow production lines to be switched with relative ease from one product line to another.

Global economic liberalization, new communications and transport technologies, and the standardization of production together provided the structural opportunities for a GPN model to emerge. The fourth factor, intense global market competition, dynamized it in the early 1990s and continues to animate it today.

Despite its high level of concentration, the global electronics industry is highly competitive. In a textbook case of oligopoly, the dominant firms maintain a fairly stable market share, and keep product prices in line with each other, either through collusion or simply following market signals. Through their market power, the dominant firms set market prices to maximize their profits. Moreover, in the absence of competition, both process and product innovation tends to be low.

Day-to-day life in the oligopolistic flagship firms is far from stable in any of these dimensions: technological innovation, product prices, or market share. A high, almost frenzied level of technological innovation and product competition drives the global electronics industry. In 1965, Gordon Moore, the co-founder of Intel, predicted that technical break-throughs in the design and manufacturing of silicon chips would enable firms to double the number of transistors on a chip roughly every 18 months. The smaller spaces between the transistors would greatly enhance the power and thus the applications of chips. Dubbed "Moore's Law," the prediction has been accurate, propelling the creation of much more powerful chips at roughly the same costs of production (Mazurek 1999).

Innovation in chip technology has been matched by innovation in applications, generating a huge array of new electronic products. As with chips, product innovations have greatly increased the power and speed of electronic products such as computers without increasing production costs. As a result, the electronics industry is characterized by rapidly falling product prices. "Disruptive technologies"—new frontier tech-nologies that fundamentally change production processes or products themselves—can trigger market realignments among the major players and create cracks for new entrants (Christensen 1997).

A key characteristic of global competition is that global markets are integrated—market position in one country affects its position elsewhere. This means that, to be competitive, the global flagship must be present in all major growth markets or risk losing global market share. Loss of market share, in turn, undermines R&D and product innovation capaci-ties—which could sound the death knell, at least for a particular product or market niche.

The IT industry is thus an *unstable* oligopoly, characterized by a high level and predatory form of global competition. According to Ernst and Kim (2001, p. 6), "mutual raiding of established market segments has become the norm," forcing firms to "engage in strategic games to pre-empt a competitor's move."

Strategic behavior among oligopolistic firms is not confined to the IT industry. Indeed, it is the signal form of competition in a global economy (Grossman 1992). However, it is particularly intense in a technologically dynamic, knowledge-intensive industry like electronics. In this industry,

Intense price competition needs to be combined with product differentiation, in a situation where continuous price wars erode profit margins. Of critical importance, however, has been speed-to-market: getting the right product to the largest volume segment of the market right on time can provide huge profits. Being late can be a disaster, and may even drive a firm out of business. The result has been an increasing uncertainty and volatility and a destabilization of established market leadership positions. (Ernst and Kim 2001)

The personal computer (PC) industry offers a case in point:

Familiar brands such as Packard Bell, DEC, AST and Micron Electronics have followed early leaders such as Osborne and Tandy into oblivion. One-time industry leader IBM has scaled back its PC business after losing billions of collars and long-time innovator Apple holds on to a shrinking niche market. Meanwhile, Dell Computer, which held less than 4 percent of the worldwide PC market as recently as 1995, is now the number one PC vendor with 13 percent of the global market in [2002] and holds nearly a quarter of the US market. (Dedrick and Kraemer 2002, p. 10)

By 2004, Dell's share had increased to more than 16 percent of global PC shipments.

In this context of extreme market volatility and price competition, flagship companies are driven to constantly seek ways to reduce cost, increase speed to market, and reduce risk. By pooling manufacturing services across multiple brand-name company clients, contract manufacturers provide all three advantages over in-house production. First, by adding orders from different clients, CMs capture economies of scale, reducing per unit cost production costs. Scale economies are generated not only on the factory floor but through the management of global supply chains for components and services. Second, generic manufacturing capacities allow CMs to rapidly reallocate production lines from one product to another, capturing economies of speed as market conditions and client orders shift. Third, by outsourcing to CMs, flagship companies avoid investment in expensive, automated in-house manufacturing capacity, thus reducing financial risk.

The GPN model feeds on itself: as productive functions become more modular, the global industry becomes larger, more complex, and more competitive, generating both new opportunities and new pressures on firms to undertake expansion through external contracting rather than by adding to in-house capacity. As Ernst and Kim argue (2001, p. 6), "no firm, not even a dominant market leader can generate all the different

Table 3.2
PC top 10 market share rankings (thousands of units).

2004 rank	2003 rank	Company	2004 shipments	2003 shipments	2004/2003 growth	2004 market share
1	1	Dell	31,340	25,600	22.8%	16.4%
2	2	HP	27,760	25,099	10.6%	14.5%
3	3	IBM	10,418	8,921	16.8%	5.4%
4	4	Fujitsu-Siemens	7,162	6,370	12.4%	3.7%
5	6	Acer	6,432	4,470	44%	3.4%
6	5	Toshiba	5,822	5,002	16.4%	3%
7	8	Lenovo	4,315	3,756	14.9%	2.3%
8	7	NEC	4,280	4,153	3.1%	2.2%
9	n.a.	Gateway	3,679	n.a.	n.a.	1.9%
10	10	Apple	3,507	3,098	13.2%	1.8%
		Others	86,609	79,277	9.2%	45.2%
		Total shipments	191,417	168,846	13.4%	100%

capabilities internally that are necessary to cope with the requirements of global competition."

Pressures for greater external contracting by the global flagships generate industry concentration in other parts of the GPN, such as contract manufacturing. When the outsourcing trend began in the late 1980s and the early 1990s, the flagships first contracted out to smaller regional contract manufacturers in Asia, Europe, and North America, including the "big five" described above. As the scale of outsourcing increased, the flagships found themselves managing an ever-expanding number and growing complexity of supplier relationships. In response, they began to consolidate manufacturing contracts with a limited number of CMs and requiring that they "go global." The industry analyst Timothy Sturgeon explains it as follows (2002):

In order to streamline the management of their outsourcing relationships, brand-name electronics firms increasingly demanded that their key contractors have a "global footprint." As a result, [what are today] the largest contractors have been aggressively internationalizing since the mid-1990s. For example . . . Solectron was concentrated in a single campus in Silicon Valley until 1991, when its key customers, including Sun Microsystems, Hewlett-Packard, and IBM, began to demand global manufacturing and process engineering support. Within ten years, the company's footprint had expanded to nearly 50 facilities worldwide.

The GPN model generates downward price pressures all along the supply chain and on to final products. The CMs compete for contracts to the brand-name companies, while local suppliers compete with countless others in multiple locations for CM customers. At the end of the process, the flagships must set prices that will not only maintain their market share but grab some of their competitor's share. Far from stable rents, flagships are vulnerable themselves to rapid erosions of profit margins.

In addition, the rapid growth and intense competitive dynamics of the industry generate endemic problems of investor overexuberance and production overcapacity, leading to crisis and shake-out. Ballooning growth in the 1990s culminated in the crash in the value of technology stocks in 2001. Revenues declined and profit margins were squeezed or reallocated to upstream suppliers such as Intel and Microsoft. Industry giants sought recovery through mergers, such as the Hewlett-Packard takeover of Compaq, as well as relocation to even cheaper manufacturing sites—especially China.

The Global Geography of IT Manufacturing

The globalization of the IT industry has propelled global dispersion of both marketing and manufacturing, and increasingly research and design functions as well. Hewlett-Packard, for example, claims to serve more than a billion customers in 178 countries on five continents. It contracts with CMs and other suppliers in 26 countries. Moreover, the company's R&D arm, HP Labs, operates in six sites around the globe, including its headquarters in Palo Alto, a substantial research facility in Bristol, and labs in Bangalore, in Cambridge, Massachusetts, in Haifa, and in Tokyo.

Despite growing global dispersion, the primary feature of the global IT industry is its tendency to concentrate manufacturing in a few countries. Both the CMs and components suppliers of the IT industry tend to be highly concentrated, with the United States and East Asia leading the pack. In the computer industry, for example, the United States accounts for about a quarter of global hardware production, while East Asia accounts for another 45 percent. While East Asian production is most concentrated in Japan, production is growing fast in China, rising from under 2 percent to nearly 7 percent of global production between 1995 and 2000.

Suppliers to the global IT industry are also highly concentrated in East Asia. Hewlett-Packard, for example, sources more than 70 percent of its product materials, components and services from ten countries in East Asia.

Global electronics *market*s are also highly concentrated, with 16 countries in North America, Europe, and East Asia accounting for nearly 65 percent of the total. Valued at US$362 billion in 2003, the United States is by far the biggest market. But China has overtaken Japan as the world's second-largest market, with sales of US$138 billion. Moreover, the Chinese electronics market is projected to grow by 13 percent a year through 2007, compared to 4 percent for the United States (Yearbook of World Electronics Data 2004).

Why is IT hardware manufacturing so concentrated in East Asia? The initial drivers, which we explore in chapter 4, were primarily far-sighted and effective government policies and institutions that generated

Table 3.3

Leading computer hardware producing countries. Source: Dedrick and Kraemer 2002.

	1995 share	Global rank	2000 share	Global rank
Americas				
US	26.5%	1	26.2%	1
Brazil	2.3%	10	2.7%	11
Mexico	1.1%	15	3.0%	9
East Asia				
Japan	25.2%	2	16.3%	2
Singapore	7.3%	3	7.6%	3
Taiwan	5.6%	4	6.5%	5
China	1.9%	12	6.8%	4
Malaysia	1.8%	14	3.1%	8
Thailand	1.9%	13	2.6%	12
South Korea	2.4%	8	2.3%	13
Total East Asia	46.1%		45.2%	
Europe				
UK	4.7%	5	4.8%	6
Germany	2.8%	6	3.5%	7
Ireland	2.2%	11	3.0%	10
France	2.7%	7	2.1%	14
Italy	2.3%	9	1.7%	15

"virtuous circles" of ever-growing local manufacturing capacities—which in turn spawned new domestic firms and attracted foreign investment. In the 1960s and the 1970s, Taiwan and South Korea were early leaders in targeting electronics in their domestic industry policies, followed in the 1980s and the 1990s by Singapore, Malaysia, and most significantly China. "China has established a critical mass of infrastructure and factory capacity that makes it the world's most important global outsourcing platform for manufacturing. It has an unparalleled mix of economies of scale, industrial diversification, and domestically funded infrastructure, buttressed by the world's largest inflows of foreign direct investment." (Gereffi and Sturgeon 2004)

Many Southeast Asian countries have undergone rapid industrial transformation in recent decades. Between 1980 and 1998, Singapore doubled the share of manufactures in its total exports from 43 percent to 86 percent, Thailand from 25 percent to 74 percent, Malaysia from 19 percent to 79 percent and Indonesia from 2 percent to 45 percent. "The only non-Asian country to undergo a transformation of a similar magnitude is Mexico, with its manufactured exports growing from just 10 percent in 1980 to 85 percent by the end of the 1990s." (ibid.)

Outside of East Asia, IT manufacturing tends to be sited in local geographical clusters—particular cities serving a particular national or regional market. In Europe, for example, there are manufacturing clusters in Ireland and Scotland, as well as Bulgaria. As this book demonstrates, the city of Guadalajara in Mexico had great promise in the 1990s as a manufacturing cluster for the US market.

The geographical clustering phenomenon is driven by what economic theory calls "economies of agglomeration," whereby firms reap benefits by co-locating with firms who provide specialized skills and services. Clustering promotes learning and sharing of ideas, which can promote innovation. Clustering also occurs when product engineering, development, and ramp-up require close interaction between engineers in assembly and supplier firms. Innovative industry segments who benefit from specialized skills, such as disk drives and flat-panel displays, tend to cluster (Dedrick 2002).

Not all segments of the IT industry, however, need specialized skills. Dedrick and Kraemer (2002) argue that the PC industry is a "mature industry" with virtually no innovation and no need for specialized skills. PC makers may still be drawn to particular countries and cities by government incentives and infrastructure. However, they predict that the need to locate near a national or regional market will become much less important in the future. PC manufacturing, they suggest, will shift from regional production to production in Asia—basically China—for global markets:

The PC industry appears to be reaching maturity, with replacement cycles slowing and most markets saturated [and] a declining number of PC makers, distributors, contract manufacturers, and suppliers of some components. . . . The result could be a two-tier production network, with complex build-to-order products being produced near the end customer and standardized products being produced *in one or two locations to supply worldwide demand.* (Dedrick and Kraemer 2002, p. 29, emphasis added)

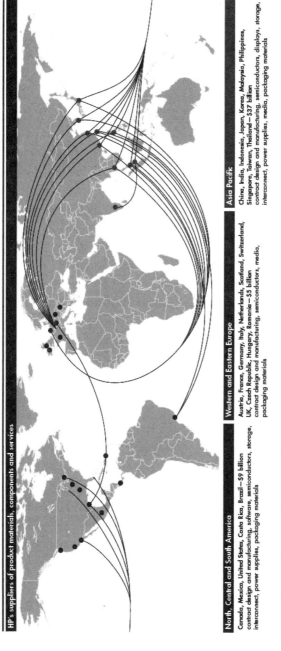

Major locations of HP product materials, components and services suppliers and key logistics routes[1]

HP's suppliers of product materials, components and services

North, Central and South America

Canada, Mexico, United States, Costa Rica, Brazil—$9 billion
contract design and manufacturing, software, semiconductors, storage,
interconnect, power supplies, packaging materials

Western and Eastern Europe

Austria, France, Germany, Italy, Netherlands, Scotland, Switzerland,
UK, Czech Republic, Hungary, Romania—$5 billion
contract design and manufacturing, semiconductors, media,
packaging materials

Asia Pacific

China, India, Indonesia, Japan, Korea, Malaysia, Philippines,
Singapore, Taiwan, Thailand—$37 billion
contract design and manufacturing, semiconductors, displays, storage,
interconnect, power supplies, media, packaging materials

NOTE: Expenditures estimated for HP FY2004 total approximately $52 billion. Total in this chart is different than $52 billion due to rounding and uncategorized spending. Locations subject to frequent change.

[1] Sites and transportation routes on the map are representative and not an exact description. Locations with the largest sourcing expenditures are shown. Actual manufacturing locations may differ in some cases. Does not include HP-owned and operated manufacturing sites (see "Operations" at www.hp.com/go/operationsreport).

Figure 3.2
Hewlett-Packard's global supply chain. Modified, with permission, from www.hp.com.

If this is indeed the trend, it is not good news for Mexico and other developing countries trying to gain a foothold and/or move up the value chain in the global IT industry. As we have noted, the global IT industry is characterized by a high degree of firm concentration at "the top," that is, at the level of the flagships and contract manufacturers, while suppliers are highly fragmented. It is also concentrated by country. On both scores, the result of global concentration, Gereffi and Sturgeon argue (2004, p. 12), is that "profits are driven down at the base of global value chains because of intense competition, leaving little money for reinvestment, innovation, or wage increases. The real opportunities to move up value chains in the global economy appear to reside in a very small number of developing countries, and with the largest of these economies (like China and India), in particular sub-national regions."

Environmental Challenges of the Global IT Industry

For most developing countries, IT investment is seen not only as a driver of industrial modernization and restructuring but also as a ticket to "clean and green" production. High-tech manufacturing and assembly seem far less polluting than traditional smokestack and resource-intensive industries such as steel production and leather tanning. For this reason, the IT sector is often a low priority for environmental regulation or monitoring.

Appearances can be deceiving. There are significant environmental and health challenges associated with IT production, including the use of a high level of toxic chemicals. Environmental risks are generally exacerbated in developing countries because of lack of infrastructure, the absence of channels for citizen oversight, and environmental regulation. In the European Union, growing citizen concern is propelling more stringent environmental regulation of industry in general and the IT industry in particular. This means that, in the short term, global markets for IT manufactures will be bifurcated by the stringency of environmental standards. In developing countries manufacturing IT products for export, local environmental and health impacts would thus depend on which markets the products were targeted for. In the longer term, environmental standards may be headed for global harmonization.

Environmental and Health Problems

The widespread social use of information and communication technologies, especially computers, has major direct and indirect impacts on environmental sustainability. Some indirect impacts are positive, for example, improved energy efficiency in "smart buildings" which use automated heating and lighting systems. Indeed, some analysts have found that, at a society-wide level, computers increase energy efficiency (Romm 1999). Computers may also reduce dependence on transport due to telecommuting. Other indirect environmental impacts are not so positive, such as the increased consumption possibilities and transport demands of internet shopping.

While there are indirect environmental benefits in the consumption phase, IT products present severe direct environmental impacts in their "front" and "back" ends, that is, the production phase and the end of product life. There are three main areas of concern:

- long-term health effects on workers and, through emissions, local communities stemming from exposure to chemicals used in the production process
- environmental and health hazards at the disposal phase stemming from hazardous materials contained in computers
- energy intensity of production.

Worker Exposure to Lead and Chemicals

At the center of every IT product is a microchip—a highly manufactured silicon wafer on which hundreds or even thousands of transistors have been etched. A microchip is like a cake made up of hundreds of layers, with icing in between each one. Making a microchip is a heavily chemical and resource-intensive process.

Beyond water and energy, microchip fabrication requires an intensive use of a wide variety of chemicals in each of the repetitive four-step process of transistor etching. These chemicals encompass cleaning solvents, acid solutions, and alcohol. A 1995 survey by the US Environmental Protection Agency listed 31 categories of chemicals used in photolithography alone, one of the last stages of manufacturing (Mazurek 1999, p. 52). Wastes from photolithography include acid fumes, organic solvent vapors, liquid organic wastes, aqueous metals, and wastewater contaminated with spent cleaning solutions.

Many of the chemicals used to make micro-chips are known to be carcinogenic or teratogenic. Most, however, have never been subject to tests. "Hundreds, even thousands, of chemicals and other materials are used to manufacture electronic products," concludes a detailed study of the environmental impacts of computers (Kuehr 2003). Moreover, the chemicals used are constantly changing, in large part because they confer competitive advantage and are "trade secrets" of a microchip manufacturer. Mazurek (1999, p. 53) concluded that "chip plants use, emit, and transport a host of constantly shifting substances that are known to be among the most toxic used in contemporary industrial production." These chemicals pose risks not only in semiconductor and circuit board production but in their own manufacture.

In the United States, the semiconductor industry has been the target of numerous lawsuits by workers alleging that chemical exposure is responsible for their high rate and rare forms of cancer, as well as high rate of miscarriage. About 250 lawsuits were filed against IBM, including by a group of workers at the company's plant in San Jose, California, who claimed that they worked with known carcinogens such as benzene, ethyl alcohol, and vinyl chloride. However, the judge ruled that evidence was insufficient. In 2001, IBM settled a lawsuit filed by two former workers who claimed that their exposure to chemicals in a New York electronics plant caused severe birth defects in their son. IBM did not admit liability and cited the potential cost of a long-running trial as the major reason for settling out of court (Sorid 2003).

Action to reduce occupational hazards from chemical exposure has been hampered by the lack of epidemiological studies—and industry resistance to undertaking them. After dragging its feet for more than 5 years, the US Semiconductor Industry Association announced in August 2004 that it would fund a worker health study to assess cancer risk among US semiconductor workers over a period of more than 30 years (US Semiconductor Industry Association 2004).

While somewhat less chemical-intensive, the component manufacturing and assembly parts of the IT product cycle expose workers to some of the same hazardous substances, including solvents. Moreover, they pose new risks to workers in the form of exposure to lead, formaldehyde, and brominated flame retardants.

Assembly involves plating copper and soldering components to the plates with lead and tin. The copper plating process emits formaldehyde while the soldering process produces lead "solder drass" that is highly contaminating. Both processes pose significant risks to workers in the production process, as well as significant disposal challenges in the form of contaminated water (Kuehr 2003).

The health impacts of exposure to lead are well-documented. Depending on the extent and duration of the exposure, they range from loss of neurological function to death. Indeed, lead exposure is a leading occupational hazard and a major source of environmental hazard for children. Even moderate lead exposure in children has been linked to a significant decrease in school performance, including lower IQ scores and hyperactive and violent behavior. Both children and adults can suffer from a range of illnesses including effects on the central nervous system, kidneys, gastrointestinal tract, and blood forming system. Lead exposure also affects the reproductive system in both men and women (Occupational Knowledge International 2005).

A newer but possibly even greater hazard in IT production is the widespread use of brominated flame retardants, such as the compound polybrominated diphenyl ether (PBDE). Workers routinely add brominated flame retardants to a wide variety of IT goods, including circuit boards and plastics in computer cabinets, to reduce flammability. The compounds are bio-accumulative—they are stored in fatty tissues and are magnified as they move up the food chain—and have been found to be rapidly rising in human breast milk in North America (Kuehr 2003, p. 52). While toxicology studies are still being undertaken, there is evidence that exposure is linked to thyroid hormone disruption, neuro-developmental deficits, and possibly cancer (ibid.).

Health risks from brominated flame retardants go beyond workers. A 2004 study in the United States by the Clean Production Action and Computer Take Back Campaign found PBDE-type compounds in dust swiped from computers in homes, schools, offices, and libraries. Indeed, every single computer that was swiped was found to have the neurotoxic chemicals in its dust.

End of Product Life: Toxic Waste

When consumers have finished with a computer—often to replace it with a newer, faster version—a computer becomes a waste product. Given the enormous growth in the computer market—the billionth personal computer was shipped in April 2002—the growing piles of computer e-waste in many parts of the world are of mind-boggling proportions (Tech-Edge 2002). In the United States, annual e-waste is estimated to be 5–7 million tons per year (NSC 1999). A lot of computers end up sitting in office cupboards and garages. About half—some 3 million tons worth in 1997 alone—end up in landfills.

In landfills, chemicals and other hazardous substances inside a computer leach into land and water. The hazardous insides of a computer include lead, arsenic, selenium, brominated flame retardants, antimony trioxide, cadmium, chromium, cobalt, and mercury (figure 3.3). In the United States, about 70 percent of the heavy metals found in landfills, including mercury and cadmium, comes from electronic products. These heavy metals can contaminate groundwater and threaten drinking water.

Figure 3.3
"The hazardous insides of a computer" (source: Roberts 2004). 1: Lead in cathode-ray tube and solder. 2: Arsenic in older cathode-ray tubes. 3: Selenium in circuit boards as power-supply rectifier. 4: Brominated flame retardants in plastic casings, cables, and circuit boards. 5: Antimony trioxide as flame retardant. 6: Cadmium in circuit boards and semiconductors. 7: Chronium in steel as corrosion protection. 8: Cobalt in steel for structure and magnetivity. 9: Mercury in switches and housing.

In 2001, California and Massachusetts banned cathode- ray tubes from municipal landfills (Raymond Communications 2003).

In an effort to reduce landfill hazards, the US EPA and municipal governments have promoted computer recycling, generating a vibrant US e-recycling industry. As of 2002, about 11 percent of US e-waste was recycled. However, according to a study undertaken by two industry watchdog groups, about 80 percent of the e-waste sent to recyclers ends up being exported to Asia. Ninety percent of exported e-waste goes to China, where computers are pulled apart by hand, useful components sold, and the rest thrown into landfills, waste dumps or just the nearest piece of empty land (BAN and SVTC 2002).

Besides the nine hazardous materials described above, people dismantling CRTs by hand will be exposed to a "witch's brew" of toxic substances: beryllium, barium, plastics, including poly-vinyl-chloride, toners, and phosphor, a chemical compound described by the US Navy as "extremely toxic." Many cases of village workers in China pulling apart computers with rudimentary tools, children playing on e-waste heaps, and families living with hazardous materials strewn around villages have been documented. Even in the much better working conditions of northern Europe, a study found that workers engaged in "demanufacturing" had elevated levels of PDBE and other flame retardants (quoted in Kuehr 2003, p. 52).

Much can be done to reduce the environmental and health risks stemming from e-waste, including better product recycling and re-use programs. But "solving" the problem will require redesigning both products and production processes to eliminate the need for harmful substances or to find more benign substitutes. Some companies are already some ways down the path of "design for environment," prompted in part by new European regulations (see below). Intel, for example, is committed to a phase-out of lead, while Hewlett-Packard has a comprehensive Design for Environment program. But there is still far to go.

Energy Intensity

The production of semiconductors is highly energy intensive. To ensure purity, the process requires large amounts of de-ionized water and air conditioning. One study found that a "fab" (chip fabrication plant) that produced 6-inch wafers used more than 2 million gallons of de-ionized water, 2.5 million cubic feet of high-purity nitrogen, and 240,000 kilowatt-hours of electricity per day (Mazurek 1999, p. 48). About half the energy is used to clean and condition the air inside the "clean rooms" where the wafers are fabricated.

Four times as much energy is needed to produce a desktop computer as to power it during its entire use phase. Researchers estimate that it takes 1.3 kilogram of fossil fuels and chemicals to produce a 2-gram memory chip (Kuehr 2003, p. 4). Indeed, energy use may account for 40 percent of total production costs in semiconductor manufacturing. High energy costs, however, mean high potential for cost savings and emissions reductions through energy efficiency. The atmospheric build-up of carbon emissions is of growing concern because of its role in climate change. According to Amory Lovins of the Rocky Mountain Institute, a 92 percent reduction in carbon emissions per microchip is currently profitable (Lovins and Lovins 2001, p. 3).

Semiconductor production is also highly water-intensive and thus sensitive to local conditions and demands. Environmental concerns about water revolve around the absolute amount required and the potential for water contamination from plant emissions. In Costa Rica, a coalition of environmental and community groups raised concerns about an Intel assembly plant that was built over several groundwater sources and near a river. Intel agreed to return wastes that could not be adequately treated in Costa Rica to the United States and worked with local environmental authorities and NGOs to implement strict water-management standards (Leighton et al. 2002).

Rising EU Standards: Globalization or Bifurcation?
For developing countries, the environmental challenges associated with the global IT industry include not only direct and indirect ecological impacts but also growing efforts to regulate and reduce them. Public concern about the environmental and health impacts of IT products and production processes has been mounting in Europe, Japan, and the United States. Environmental and municipal authorities, as well as companies, are undertaking a wide range of regulatory and voluntary initiatives to reduce chemical hazards and electronic waste and to improve energy and resource efficiency.

The European Union, however, has gone the farthest toward "raising the bar" in terms of environmental regulation of the IT industry, as well as hazardous chemicals in general. A signatory to Annex One of the Basel Convention, which prohibits the export of hazardous waste from developed countries, the EU has focused regulation on reducing the amount of waste entering, produced, and stored in Europe.

The issue of e-waste is very much in the public eye in Europe. In 2005, the Royal Society of Arts in the United Kingdom built an "e-waste man" made up of the average quantity of electrical and electronic wastes a European disposes of in a lifetime, including five refrigerators, 12 kettles, and 35 mobile phones. The sculpture, which toured the United Kingdom, weighed 3 tonnes and stood 7 meters tall.

Along with new sweeping regulation of chemicals, two EU directives will change the rules for IT market access in Europe:

Restriction on Hazardous Substances (ROHS)   The RoHS Directive will ban the placing on the EU market of new electrical and electronic equipment containing more than agreed levels of lead, cadmium, mercury, hexavalent chromium, and the flame retardants polybrominated biphenyl (PBB) and polybrominated diphenyl ether (PBDE). There are a number of exempted applications for these substances. Manufacturers will need to ensure that their products—and their components—comply in order to stay on the Single Market. If they do not, they will need to redesign products. The directive went into effect July 1, 2006.

Directive on Waste from Electrical and Electronic Equipment (WEEE) Adopted in the spring of 2003, WEEE aims to increase the recycling and recovery of electronic waste in Europe by making the manufacturers of electronic goods assume responsibility for their products. It applies to a huge spectrum of products. It sets criteria for the collection, treatment, recycling, and recovery of waste from electrical and electronic equipment and makes producers responsible for financing most of these activities (producer responsibility). Private householders are to be able to return electronic waste without charge.

Both OEMs and CMs are working to redesign products and production processes in order to meet the new EU requirements, including in China. The EHS manager of Flextronics, Mexico, for example, told us that his CEO is sending out the message that "business is changing" and the company's success depends on adaptation to environmental conditions and demands, especially Europe's "take back" legislation and regulation of heavy metals. Under the WEEE initiative, he stressed, companies will own products, while consumers will simply pay for use rights (Ochoa 2003).

Other large electronic markets, notably the United States and Japan, have not adopted such laws, at least not yet. In recent years, Europe has emerged as a leader on a wide range of environmental and consumer protection initiatives, ranging from greenhouse gas reduction to corporate social responsibility. While the United States and Japan are taking both regulatory and voluntary actions of their own, they are also watching Europe. In the United States, for example, the EPA is working with chemical manufacturers to phase out the most toxic of the brominated flame retardants and Congressional legislation was introduced to ban them permanently (Pohl 2004).

In the context of globalization, MNCs may find it too costly to design products and processes around multiple standards. "The EU is going where no man has gone before," says James Lovegrove, managing director of the European division of the American Electronics Association, a US industry lobby. "The moment the ink hits the paper in Europe it becomes a global piece of legislation." (quoted in Pohl 2004) Indeed, some IT leaders, like Tim Mohin, Sustainable Development Director for Intel, would like to speed up the process and advocates global harmonization of environmental, health and safety standards. According to Mohin, global harmonization would provide both social benefits and a boost to company bottom lines (Mohin 2005).

Global harmonization of regulatory standards may be in the cards in the future. In the short to medium term, however, IT markets will be at least somewhat bifurcated along the lines of higher EU versus lower US environmental and health standards. Japan, which was the first country to pass a law requiring recycling of domestic e-products, is likely to be in between. While there is mounting concern about landfill contamination from e-waste, the United States is not a signatory to Annex One of the Basel Convention. Indeed, the US EPA views the export of e-waste to Asia as part of its industrial waste management strategy (Raymond Communications 2003).

In global markets bifurcated by environmental demands, developing countries—and the MNCs that invest in them—will need to strategically choose and align their environmental policies with their markets. If they choose markets with lower standards, however, they may be at risk of being squeezed out if and when standards are globally harmonized.

4

Wired for Sustainable Development? IT and Late Industrialization

Developing countries are eager to gain or upgrade their place in the global IT industry. A handful—almost exclusively in Asia—have done so with significant success. While different in detail, the success of Asian countries in capturing development benefits from foreign IT firms was due to two shared features: (1) a match between a country's location-specific socio-economic and/or geographical assets and global production strategies of MNCs and (2) a match between national linkage capacity, including government policy in nurturing local firms and MNCs' interest in local sourcing.

Promises and Pitfalls of IT Investment

Foreign investment in high-tech industries in general and the IT industry in particular is very attractive to developing countries. Beyond immediate benefits common to all FDI, such as jobs and foreign exchange, high-tech FDI offers the promise of new technology, new skills, and a link with the world's fastest-growing industry. As Alice Amsden explains (2004, p. 87), "a high tech industry is one whose technology is still tacit rather than explicit owing to firm-specific, proprietary capabilities that create novel products and earn above-normal rents. High-tech industries are thus desirable for a country because they require high-wage, skilled workers and offer opportunities for entrepreneurs to earn technological profits."

In most developing countries, domestic firms lack the manufacturing and innovation capacities to enter any but the lowest rungs of global high-tech production as competitors. Moreover, as outlined above, there are very high barriers to entry in the IT industry for would-be flagships, contract manufacturers, and even higher-level suppliers. "The domination of the United States and Asia in electronics," Amsden argues, "makes it very difficult for newcomers to enter this field." Even in East Asia, it is not clear that important producers like Malaysia and Singapore will be able to meet the challenge of continuous upgrading in the face of the intense competitive dynamics of the industry itself, as well as new competition from China and India (Ernst 2003; Wong 2003).

The primary route for developing countries trying to build to a local IT industry has been to start off by attracting FDI in low-wage export platforms for low-skilled and semi-skilled manufacturing and assembly operations. The plants are owned and operated by foreign CMs and higher-tier suppliers and, in the main, products are exported to markets in the United States, in Europe, or in Japan.

Even though production is highly standardized and routinized, the hope is that knowledge spill-overs will nurture the innovation and manufacturing capacities of domestic firms. Eventually, these firms will be able to upgrade technology and skills, enabling them to compete globally in more complex, higher-value-added parts of the IT value chain, as well as in niche markets. "Technological upgrading is facilitated by entering into high-tech global value chains, even at the assembly level (for export-oriented operations)," the UN Industrial Development Organization counsels in a detailed report examining the role of innovation in economic development (UNIDO 2002, p. 10). Ernst and Kim concur that global production networks "have acted as a catalyst for international knowledge diffusion, providing new opportunities for local capability formation in lower-cost locations outside the industrial heartlands of North America, Western Europe and Japan" (2001, p. 2).

IT investment also offers the promise of forward linkages, that is, the diffusion of IT products in the domestic economy, with potentially enormous benefits for productivity. The UN estimates that in 2001 more information could be transmitted through a single cable in a second than was sent over the entire internet in one month in 1997. During the "roaring" 1990s, one-fourth of global economic growth was due to IT use and its forward linkages (UNDP 2001). Developing countries know what side of the "digital divide" they want to be on.

There is also potential for IT industrial development to bring benefits to the poor. Access to the internet and other communications technologies can expand the realm of political participation and provide greater transparency. Information access can also directly improve economic and social welfare. Information about price levels and fluctuations, for example, can help farmers to improve their negotiation positions, while fishermen can use remote sensing information to identify schools of fish. Health care officials can diagnose remote patients via web-based networks (UNDP 2001).

Though the promise is palpable, foreign IT investment may not live up to its potential. There are four pitfalls.

First, there is intense competition for IT investment, both among countries and municipalities. Brand-name flagship firms and CMs looking for a manufacturing location can pick and choose. Many developing countries simply do not have the requisite infrastructure, skills, and large domestic markets to successfully attract high-tech FDI in the first place. Trying can impose opportunity costs.

Second, intense cost pressures inside the industry undermine the sustainability of low-wage assembly work. CM assembly operations tend to be footloose, rapidly relocating operations to lower-cost locations when global conditions change—as many laid-off workers in Guadalajara found to their dismay in 2001. (See chapter 5.) Even China, which successfully leverages its cheap labor and domestic market access to build a burgeoning CM industry, is vulnerable. Asked by one of the authors how much of his company's southern China operations would relocate if land and infrastructure subsidies were withdrawn, the head of Asia operations for Flextronics responded "about 50 percent." Where to? Vietnam.

Third, there is little skill acquisition and hence few human capital spill-overs in assembly work. Because an initial level of training is required, assembly workers have been called "semi-skilled." But there is no chance to get further training and little opportunity for workers to innovate on the job. Organized into work groups, workers undertake highly standardized and repetitive tasks. Indeed, global standardization and uniformity of work procedures is a defining feature of CMs, which "offer a uniform interface for flagships seeking global one-stop shopping for manufacturing services" (Luthje 2003, p. 9).

Fourth, FDI by foreign IT firms may not generate sufficient knowledge spill-overs to promote technological upgrading. Unlike licensing arrangements, FDI does not necessarily or directly transfer technology. Besides human capital, spill-overs are indirectly captured through backward and forward linkages. But CMs draw inputs from a global supply chain and may have few backward linkages to local firms. Moreover, a neo-liberal "hands-off" policy framework inhibits governments from helping local firms develop capacities to become suppliers.

A substantial level of infrastructure, manufacturing and innovation capabilities must be nurtured by firms and the government in order to match between national linkage capability and MNC interest in sourcing from domestic suppliers. Few developing countries have such "absorptive capacities." Indeed, the focus of the UNIDO report quoted above is precisely "to determine why many developing countries are unable to use new industrial technologies efficiently" (UNIDO 2002, p. 10). The answer, in a nutshell, is the lack of institutions needed to nurture local skill acquisition and manufacturing capabilities.

Spill-overs will also be scarce if forward linkages are weak. But a single-minded focus on production for export inhibits the growth of a domestic market for IT goods and the benefits of IT domestic diffusion, including the spur to product innovation. The lack of a strong domestic market, in turn, makes it more likely that flagships and CMs will be footloose—and less likely that they will be willing to partner with local governments and firms to transfer technology and know-how.

Asian Success Stories

Starting in the 1960s, a number of countries in East Asia—notably Taiwan and Korea—leapfrogged into central places in the global IT industry. More recently, China and India have emerged as major players. Malaysia as well has made major headway but is vulnerable to backsliding. Given the industry's high barriers to entry, how did these "latecomer" nations get a seat at the table?

Mario Cimoli (2000) describes an "optimal cycle of globalization" as the way developing countries successfully ratchet up productive capacities. First, foreign firms come seeking lower labor costs and access

Table 4.1
Asia's share of developed countries (EU, Japan, US) high-tech markets. Source: United Nations 2006.

Computers and peripherals

	1981		2004	
	Trade value (US millions)	Share	Trade value (US millions)	Share
China	931,498	0.00%	75,507,781,121	26.40%
India	12,205,450	0.06%	96,506,633	0.03%
Malaysia	15,326,593	0.08%	16,563,137,584	5.79%
Taiwan	179,232,476	0.89%	18,387,518,714	6.43%
Republic of Korea	81,880,515	0.41%	10,857,058,761	3.80%
Total	289,576,532	1.44%	121,412,002,813	42.46%

Telecommunications

	1981		2004	
	Trade value (US millions)	Share	Trade value (US millions)	Share
China	6,797,935	0.03%	52,281,865,594	22.13%
India	5,685,519	0.03%	178,398,302	0.08%
Malaysia	139,929,364	0.64%	10,746,064,083	4.55%
Taiwan	1,686,482,952	7.73%	7,524,774,204	3.19%
Republic of Korea	909,972,921	4.17%	20,121,437,739	8.52%
Total	2,748,868,691	12.60%	90,852,539,922	38.46%

to national or regional markets. These firms create subsidiaries and develop local suppliers that learn how to manufacture and process at a higher level of technological sophistication. In the next stage, foreign and domestic firms may establish joint R&D activities with local universities. The hope is that out of a newly vibrant set of domestic firms, new domestic firm leaders will emerge that can export and invest abroad in their own right.

The Asian "success stories" have generally followed such a path. In 1981, the combined exports of China, India, Malaysia, Taiwan, and South Korea amount to less than 1.5 percent of all US, Japanese, and EU imports of computers and peripherals. By 2004, their share exceeded 40 percent. China's share alone went from zero to more than 26 percent. The story was similar for telecommunications products: from 12.6 percent in 1981, the share of Asian nations in rich country markets exceeded 38 percent in 2002 (table 4.1).

Taiwan and Malaysia followed Cimoli's cycle to the letter, relying heavily on FDI, though each worked differently with MNCs. Korea took a slightly different path, relying more on domestic firms at the outset, while India first nurtured state-owned enterprises. China is following a similar path today, although China is proceeding on both stages at once within its own overarching strategic framework. In all cases, the key to success was active and effective leadership by the government—not only to attract MNCs but also to nurture domestic capacities to upgrade and innovate. East Asian governments acted to "speed up" high-tech development by setting clear objectives and creating institutions and implementing strategic policies to achieve them.

Taiwan

Taiwan was the first developing country to crack into IT production for global markets. Taiwanese firms started offering a low-wage export platform for final assembly operations and then moved into contract manufacturing. Starting in the 1960s, the overarching goal of the Taiwanese government in the electronics sector was "to create local growth opportunities and local value" (Amsden 2004, p. 80). The objective was to capitalize on "second-mover advantages" to create national firms who could compete globally, first as CMs and then as flagships.

In broad terms, the strategy had three prongs: (1) leveraging Taiwan's geo-strategic role and local manufacturing capacities to attract foreign electronics firms, especially from the United States, (2) creating joint ventures and other partnerships with the flagship MNCs to gain access to technology and know-how, and (3) actively and consistently supporting national firms to develop the capacities required to be globally competitive on their own. The strategy worked, and the Taiwanese firm Acer is now among the top 20 computer manufacturers in the world (Amsden and Chu 2003).

The active support of the Taiwanese government in the form of investment and policy innovation was decisive. Taiwan made four strategic investments: (1) in large plants to generate economies of scale, (2) in higher education, (3) in science parks and R&D institutions, most notably the Industrial Technology Research Institute (ITRI), and (4) in distribution networks. These investments were a combined effort of government and industry. From the 1950s through the 1990s, the government accounted for half of all IT investment. Through "state-led networking," Taiwan entered the global market by becoming an essential supplier to and producer for the large firms in the core companies. Building on that experience as a supplier, the Taiwanese state enabled the private sector to leapfrog into the list of core companies.

The government supported—and continues to support—its strategic investments with import-substitution policies that aim to "create the new market segments in which national companies could then compete" (Amsden 2004, p. 80). In the first phase, these policies included tariff protection, local content regulations and development banking. In the second phase, favored policies were spin-offs from state-owned research institutes and science parks, subsidies to public and private R&D, tax breaks and subsidies to residents of science parks. According to Amsden (ibid.), the strong focus on R&D support paid off handsomely: "By 2000, there were more than 15,000 professionals in Taiwan who had at one time or another worked for ITRI, the government's premier research center devoted to high-tech industry. Of these . . . more than 12,000 had, in fact, gone to work in such industry. . . . ITRI was responsible for spinning off the two pillars of Taiwan's semiconductor industry, the United Microelectronics Corporation (UMC) and the Taiwan Semiconductor Manufacturing Company (TSMC)."

Korea

Like Taiwan's, Korea's leap into the global IT industry was propelled by government vision and policy while being in line with MNC strategic interests. The leap was part of Korea's larger, remarkable industrial transformation after World War II. "No country has come as far and as fast, from handicrafts to heavy industry and from poverty to prosperity, as the Republic of Korea," Linsu Kim asserts (2003).

Colonized by Japan as a site for agriculture and simple industry and torn apart by the Korea War, South Korea in 1960 looked much like many of the least developed countries today. In 1960, complex manu-factures accounted for less than 1 percent of Korea's exports. By 1999, they had risen to 63 percent, with high-tech products accounting for half of complex manufacturing exports (Kim 2003). However, unlike most developing countries today, including Mexico, Korea did not seek tech-nology transfer by encouraging unrestricted FDI. Indeed, Korea took the opposite tack—it restricted FDI whenever possible. Rather than foreign companies, the primary agents of technological upgrading were domestic firms.

Korea's strategy was to actively promote the learning capacities of domestic firms, including engineering, research and development, and global networking and marketing. Given their starting point, domestic firms had to depend on access to foreign technology and know how. Eschewing FDI and the economic control it gave to foreign investors, the government encouraged technology transfer through licensing agree-ments, reverse engineering, the import of capital goods and flagship con-tracts (that is, domestic firms subcontracting with the flagships).

Although restrictions were gradually loosened, the role of FDI remained relatively small well into the 1990s. Between 1997 and 1999, FDI inflows totaled $3.2 billion, while the value of capital goods imports totaled $40.5 billion (Kim 2003, table 6.2). Moreover, by the late 1990s, foreign investors were seeking to tap not low-wage labor but Korea's leading-edge technologies.

The government used a variety of policies to attract FDI and create linkages. Among them were to support the inflow of foreign equipment, including an overvaluation of the local currency, tariff exemptions on imported capital, and credit subsidies. There were also import restric-tions, though these were lifted by 1987, just 7 years after domestic firms began producing personal computers (Evans 1995).

In what Evans (ibid.) describes as "high technology husbandry," a host of initiatives also directly nurtured the learning and technological capacities of local firms, including financial support for government research institutes and for research in universities. For example, the Korea Advanced Institute of Science and Technology, founded by the government in 1975, trained a large number of scientists and engineers. In 1989, the government explicitly targeted basic research as one of the nation's top technological priorities and began to establish Science Research Centers and Engineering Research Centres in leading universities. These initiatives were paralleled by substantial support for primary, secondary and tertiary education.

The government also worked to stimulate and coordinate research and development efforts by the *chaebol,* Korea's large domestic firms. For example, in the 1980s, the Electronics and Telecommunications Research Institute (ETRI) worked with Samsung and other firms to develop a 4-megabit DRAM, a project explicitly aimed to boost Korea's competitiveness relative to Japan. Beyond providing small loans, the government organized and managed the large and ultimately successful research effort that was undertaken by private firms (Evans 1995). Such efforts undergirded the factor most responsible for Korea's meteoric emergence as an industrial and IT powerhouse: "rapid technological learning by domestic firms" (Kim 2003, p. 145).

India

While Taiwan and South Korea are often pointed to as success stories, some analysts in the past have wondered if the door shut after they entered the global IT market. More recently, China and India have burst in at a furious pace. Indeed, the emergence of China and India has generated both celebration and consternation, especially in the United States. Almost daily, major newspaper carry stories of telecom and electronics success in Bangalore and Beijing—and fears that they are taking away US jobs.

India singled out electronics, including IT, as a key industry for its overall development strategy nearly 30 years ago. Over that time the government orchestrated a series of efforts that, along with private sector development, created globally competitive skills and firms in key segments of the global IT market (Evans 1995; Kumar and Joseph 2004). The key initiatives were the creation of a national network of science and technology institutions; subsidies to private IT companies to help develop and diffuse technologies; and support for the building of technical skills in higher education.

Unlike the governments of Taiwan and Korea, which worked to develop private sector firms, India had a strong emphasis on the development of state-owned firms. Since 1990, however, market liberalization, including tariff reduction and 100 percent ownership by foreign firms, has increased the role of FDI and private domestic firms in both the software and hardware parts of India's IT sector. Since 1995, India's IT sector has grown by more 45 percent per year. By the turn of the century, India had emerged as a global leader, especially in software and IT services. Exports of IT go to 133 countries and now account for one-fifth of India's exports (Kumar and Joseph 2004).

Under both more liberal and less liberal FDI regimes, the active support of the government in building a science and technology skill base has been crucial to the success of India's IT sector. The government established science and R&D centers, and in 1972 established a Department of Electronics. That department set up Regional Computer Centers run like public utilities, attached to educational institutions. Since the late 1980s, the Department of Electronics has concentrated on providing R&D, data communication and networking infrastructure to the educational and research community and to the software industry. Another institutional invention was the establishment of Software Technology Parks to provide the necessary infrastructure for software export.

Besides institution-building, the government supported the industry through subsidies, including procurement policies, Export-Import Bank loans to private companies, and support for market research in industry (Kumar and Joseph 2004). In addition, information is easily transmitted via satellite links paid for by the government (Balasubramanyam 1997).

A third area of strong Indian government support for IT is education and training. India developed specialized university programs, technical institutes, and other programs aimed at building a skilled workforce for information and communications industries. Proficiency in computer programming is mandatory for undergraduates and science post graduates in all major universities. The Department of Electronics set up the Computer Manpower Development Program at 400 institutions of learning that produced 150,000 personnel by 1996. In the private sector, private firms successfully lobbied the US government who made it easier for Indians to receive visas for attendance and teaching at institutions of higher learning in the United States (Balasubramanyam 1997; Arora et al. 2001; Kumar and Joseph 2004).

One of India's location-specific assets relative to other developing countries is its English-speaking low-wage work force. In the 1990s, India became a magnet for the outsourcing of phone services and technical support parts of the IT services industry (Kumar 2004). Moreover, India trained many more IT professionals than were able to find employment at home. An exodus of Indian engineers found their way to IT firms in the United States. Along with other parts of the large Indian diaspora, these engineers heavily supported the sector in the form of remittances and investments. When the dot com boom ended in 2001, many returned to India with contacts with US firms and a knowledge of business know-how and industry culture (Balasubramanyam 1997; Arora 2001). Combined with India's strong science and technology base, these factors suggest that India is primed for very strong IT growth in coming decades.

Malaysia

In the 1980s and the 1990s, Malaysia was transformed from a primarily agricultural and resource-based to a manufacturing economy. At the center was electronics. From less than 1 percent in 1968, electrical and electronic products rose to 71 percent of Malaysia's manufactured exports in 1997 (Rasiah 2003, p. 309). By the mid 1990s, Malaysia was a major player in global electronics and IT markets. By 2004, Malaysia's shares in developed imports of computers and telecommunications had surpassed that of Taiwan and Korea respectively (table 4.1). While impressive, these figures mask the fact that the *composition* of Malaysian exports has changed very little. Unlike Taiwan and China that have moved up the value chain, Malaysia—like Mexico—remains largely a final assembly export platform.

Malaysia's strategy was to rely heavily on relatively un-assisted foreign direct investment. Indeed, with the exception of Singapore, Malaysia is far more exposed to and integrated in the global production networks of the global IT industry than other East Asian producers. According to Ernst (2003, p. 13), "Malaysia's electronics industry continues to be shaped by strategic decisions of global flagships (both flagships and major American CMs)."

Until the mid 1990s, the strategy generated significant employment and productivity growth. Since the mid 1990s, however, productivity growth has slowed and layoffs have increased. While Malaysian firms are adept at changing product lines, they have not moved much beyond generic manufacturing capacities into higher value-added activities. "Structural change in products," Rasiah (2003, p. 312) concludes, "has not been matched by a similar upgrading in *functions*" and Malaysia's FDI-driven economy "remains specialized in low value-added functions within high-tech activities." As a result, Malaysia's electronics sector is vulnerable to the relocation of flagships and CMs to countries with lower labor costs, especially China.

What accounts for Malaysia's failure to date to upgrade its industrial capabilities? Recent studies point to two shortcomings. First, there is a serious shortage of specialized skilled labor in Malaysia, including managerial, scientific and engineering. Despite a variety of government incentives to MNCs to stay or locate in Malaysia, especially Penang, lack of sufficient investment in education and training has created a "human resource bottleneck" (Ernst 2003, p. 17). As a result, "growing deficits in skill and innovation are weakening the foundations of long-term growth" (Rasiah 2003). Second, Malaysia's export-oriented electronics sector is highly import-dependent: some 43 percent of intermediate goods are imported. Unlike Taiwan, Korea and Singapore, Malaysia "has failed to develop a broad and multi-tier base of support industries" (Ernst 2003, p. 14). One reason may be the large role of CMs, who tend to keep their design functions in the United States and Europe and thus create "only limited upgrading opportunities, insufficient for a major push into more knowledge-intensive activities" (ibid).

More fundamental is the lack of proactive government policies. The key difference between Korea and Taiwan and Malaysia, Rasiah concludes, is that "the government did not undertake policies to build local innovative capabilities" (Rasiah 2003, p. 320).

China

The most impressive contemporary IT leapfrogger is China. For 25 years, China gradually and quietly built manufacturing capacities and integrated into world markets. China is now at the core of MNC location strategy because of its multiple location-specific assets: a large and growing internal market *and* a low-cost export platform for manufactured goods.

Furthermore, China provides a match between national linkage capability between MNCs and domestic suppliers.

Much is made of China's low wages as a major factor driving MNC outsourcing to China and IT development more generally. There is little doubt that wages are low: the average manufacturing wage in China was estimated to be 61 cents an hour in 2001, compared to $16.14 in the United States and $2.08 in Mexico (Federal Reserve Bank of Dallas 2003). But the story of China's success and likely emergence as the center for global IT production goes beyond low wages and generic product manufacturing capabilities.

In 1986, four Chinese scientists recommended to the government that IT be designated a strategic sector. The request was approved and in 1988, China's National Development and Reform Commission (formerly the State Planning Commission) designated high tech as a "pillar" industry worthy of strategic industrial policy (MOST 2006). It was coupled with the Ministry of Science and Technology's (MOST) National High Tech R&D (or 863) Program that supported R&D efforts of local governments, national firms, and regions.

The goal was to foster a vibrant high-tech sector with national firms that could eventually compete as global flagships, CMs, and suppliers (Dussel 2005). The strategy was to establish domestic firms and bring foreign firms to China to build their capacity to produce components and peripherals for PCs. To this end, IBM, HP, Toshiba, and Compaq were all invited to come to China and form joint ventures with such Chinese firms as Legend, Great Wall, Trontru, and Star. China required the foreign firms to transfer specific technologies to the joint venture, establish R&D centers, source to local firms, and train Chinese employees (USDOC 2006). By the 1990s, all of the major CMs also came to China under similar arrangements.

The strategy paid off handsomely. "By carefully nurturing its domestic computing industry through tightly controlled partnerships with foreign manufacturers," Dedrick concludes (2002, p. 28), "China has become the fourth-largest computer maker in the world."

Why did the MNC flagships accept these deals? First, China had location-specific assets that could not be ignored. Not only did China offer an export platform like Taiwan and South Korea did, but they also had a large and growing market at home. In essence, foreign firms traded market access for technology transfer.

A major bargaining chip, China's domestic market is growing rapidly, propelled not only by a rise in personal income but by active government promotion strategies. In 1990 there were only 500,000 personal computers in China (ibid.). By 2000, 7 million PCs were sold in China in just one year, and by 2003 annual sales had risen to 13.3 million (CalTrade Report 2004). While the United States remains the world's largest market, access to China's domestic market is essential to the future global competitiveness of all major IT companies. Edward Barnholt, CEO of Agilent Technologies, made the point at a conference on Global Business at Stanford University in 2004 (Barnholt 2004): "[As] more and more people come into the economy into China and India, there is [*sic*] going to be huge opportunities for companies. In fact from a consumer market point-of-view, for companies like HP that is the biggest market opportunity."

Despite the potential market payoffs, foreign firms are starting to get nervous about technology transfer arrangements, especially as Chinese IT firms begin to emerge as flagships. Indeed, OECD governments have begun to dub China's policies as "forced transfers" and have undertaken investigations and task forces in order to eliminate or reduce them (USDOC 2006). In bilateral and regional trade deals with investment chapters, the US government has seen to it that such transfers are illegal.

In addition to domestic market access, global MNCs are willing—indeed, eager—to work in the confines of Chinese policy because of China's active support for and subsidies to the high-tech industry. According to a comprehensive study by Dussel (2005), a key program has been the establishment of high-tech industrial parks. There are now 25 high-tech parks in five cities: Beijing, Tianjin, Shanghai, Shenzen, and Suzhou. The parks receive subsidies in the form of tax breaks and low costs of infrastructure, including energy and water. Much of the FDI flows to these parks where it is matched with national firms who are the recipients of numerous incentives and assistance programs.

China's high-tech promotion strategy, in short, had two prongs: stimulating investment and technology transfer by MNCs and building

Table 4.2
High-tech exports and ownership in China, 1993–2003. Source: Dussel 2005.

	Billions of dollars		Percentage		
	1993	2003	1993	2003	Growth rate
Computers and peripherals	0.7	41	100	100	50.2
Collective	0	0.4	0	1	—
State-owned enterprises	0.2	2.5	26	6	29.7
Joint production	0	0.8	4	2	40.2
Joint investment	0.1	6.2	19	15	46.7
100% foreign owned	0.4	30.8	51	75	56.1
Other	0	0.4	0	1	—
Telecommunications and equipment	12.3	89	100	100	21.9
Collective	0.1	2.7	1	3	36
State-owned enterprises	6.6	16	54	18	9.2
Joint production	0.9	2.7	7	3	12
Joint investment	2.8	24.9	23	28	24.3
100% foreign owned	1.8	38.3	15	43	35.4
Other	0	4.5	0	5	—

up domestic firms. To provide investment for domestic firms, China established two funds: the Export Development Fund for the larger firms and the Fund for Small and Medium Enterprise Incursions into International Markets for suppliers. The government also offered value-added tax refunds to exporting firms, and the Chinese Export-Import Bank provided loans at preferential interest rates.

Another key element of the strategy is a high level of support for high-tech R&D and education. According to MOST, the national government spent 1.23 percent of its GDP on R&D activities in 2005, a large part in IT. R&D funds are distributed to state-owned enterprises, local governments, and Chinese-owned firms. The 2004–2008 five-year plan calls for increased subsidies to SOEs (table 4.2). Support for local government is targeted at the cities which house R&D centers within industrial parks. Local governments often match national government funding for R&D programs. In tandem with R&D, China has a high level of support for tertiary education: more than 20,000 scientists and engineers graduate from Chinese universities per year (MOST 2006).

The results of China's high-tech program have been impressive. By 1989, the Legend group had evolved into Legend Computer and formed a joint venture with Hewlett-Packard. By 2000, Legend had emerged as the number one seller of personal computers in Asia Pacific and held more than 20 percent of the Chinese PC market. In early 2005, Legend (now Lenovo) acquired IBM's global desktop and notebook computer divisions. With the IBM deal, Lenovo became, after Dell and HP, the world third-largest PC maker (Spooner 2005).

By 2003, China's electronics sector generated $142 billion in exports and employed 4 million workers. Between 1993 and 2003, the growth rate of high-tech exports was 50.2 percent for computers and peripherals, and 21.9 percent for telecommunications and related equipment (table 4.2). Like Lenovo, many Chinese firms started as state-owned enterprises and were gradually privatized as they gained capacity and competitiveness. In 1993, 26 percent of computer and peripheral firms and 54 percent of telecommunications firms were SOEs. By 2003, only 6 percent of computer and 18 percent of telecom firms were SOEs.

Although national firms and SOEs are in the minority, they are filing and being granted more patent applications. According to MOST, Chinese firms were granted 112,103 patents in 2002, whereas foreign firms were granted only 20,296. Nearly half of these patents were in the form of utility models—patents for incremental innovations where local firms create variations on project and process execution. This reveals that a significant amount of learning is going on in Chinese firms. Another half however, is in the form of design patents, of which 49,143 were awarded in 2002 (MOST 2006).

China recognizes that exporting IT goods is essential to technological development. By selling on world markets they must meet or beat global technological standards. That is why exporting is encouraged through so many policies in China. The culmination of these policies of course was China's entry into the World Trade Organization in 2001. After less than 13 years of fostering an endogenous industry, China's IT and electronics firms can now compete in world markets. Furthermore, relative to bilateral trade deals such as NAFTA, the WTO offers China much more room to maneuver to foster industry and create avenues for learning, avenues that China is exploiting to the max. Whereas NAFTA's investment chapter makes it difficult for nations to require joint ventures with R&D and learning standards, the WTO still leaves China the policy space to do so.

Environmental Legacy

Asia's brilliant success in launching national manufacturing industries, including IT, came at a colossal environmental cost. Attributed to a philosophy of "grow now and clean up later," average levels of air pollution in the 1990s were 5 times those of OECD countries and twice the world average. Water pollution fares no better, with biochemical oxygen demand and suspended solid levels well above world averages (table 4.3).

Environmental hazards in the IT sector are substantially documented. (See chapter 3.) With its witch's brew of lead, heavy metals, and toxic chemicals, the production and disposal of IT products pose hazards to workers and the environment wherever they occur. But there are extra hazards in developing countries stemming from lack of infrastructure and regulatory oversight. In the Philippines, for example, an environmental law mandates that toxic waste not be stored on sight. Complying with the law, IT companies hire trucks to haul it away. But there are no toxic waste treatment facilities in the country (although there is now a regional facility in Malaysia). Where, then, does the waste go?

Besides lack of infrastructure, developing countries lack capacities—and often political will—for regulatory oversight. The scarce resources of environment ministries prioritize "dirtier" industries, leaving high-tech companies (including foreign companies) to self-regulate. While much

Table 4.3
Environmental conditions in Asia in the 1990s. Source: Angel et al. 2000.

	Asia	Africa	Latin America	OECD	World
Air pollution					
Particulates (mg/m^3)	248	29	40	49	126
SO$_2$ (mg/m^3)	0.023	0.015	0.014	0.068	0.059
Water pollution					
Suspended solids (mg/l)	638	224	97	20	151
Biochemical oxygen demand (mg/l)	4.8	4.3	1.6	3.2	3.5

has been made of the tendency for multinationals to implement global standards, in reality MNCs have often followed local standards, or have selectively implemented global standards. (See chapter 1.)

Occupational health standards in particular tend to be sketchy or non-existent in developing countries, and fall through the cracks in the global MNC standards. In Thailand, for example, occupational health problems became visible in the early 1990s when four workers at a Seagate disk drive facility died after a pattern of fatigue and fainting. A study by the country's most prominent occupational health doctor found that some 200 plant employees had blood levels that suggested chronic lead poisoning, perhaps aggravated by exposure to solvents. In response, Seagate pressured the Thai government to prohibit the doctor from practicing occupational medicine. Thai NGOs, on the other hand, mobilized to pressure the government to enact worker protection laws (Foran 2001).

Even Taiwan (a relatively well-off industrializing country) has suffered from the effects of toxic waste. In a study of the Hsinchu Science-Based Industrial Park, the heart of Taiwan's high-tech industry, the Taiwan Environmental Action Network found a shocking level of neglect of risks to human health and the environment, starting from the 1960s and continuing through to 2001. The legacy of Taiwan's spectacularly successful high-tech development is a cohort of former employees with a high rate of rare cancers, and a severe and widespread problem of freshwater and coastal pollution (Shang et al. 2003).

In Malaysia, the disposal of toxic high-tech waste has posed a serious environmental problem. In the 1990s, the Kualiti Alam waste treatment facility came on line. However, surveys concluded that only major companies would utilize the facility, due to the lack of enforcement of environmental regulations, as well as resistance to any added cost by small and medium-size companies. After a lengthy dispute with industry, a new pricing system reduced the cost of treatment. But the incinerator capacity is only 100 tons per day, while as much as 300 tons per day arrive on site. The remainder likely finds its way into the local environment (Zarsky et al. 2002).

Rock and Angel (2005) found that some MNCs are working to clean up their environmental act worldwide and in Asia in particular. In a comprehensive study of environmental policy for the Motorola Company in Penang, Malaysia, they found that Motorola Penang not only had relatively comprehensive environmental standards for its plant but also made significant requirements of its suppliers. In the 1990s Motorola adopted a set of global firm-based environmental standards that focus on performance, procedures and suppliers. A key incentive for introducing these policies was a perception of the need to comply with the RoHS and WEEE initiatives in the EU that were described in chapter 3 above (Rock and Angel 2005).

In the years 2000–2003, according to Rock and Angel, Motorola's Penang plant reduced its use of hazardous materials by 75 percent, increased its scrap recycling by 100 percent, and reduced its water use by 42 percent. For its suppliers, Motorola mandated that all first tier suppliers phase out the use of CFCs and reduce the use of lead and other metal toxics. In the process, with one of its chemical suppliers Motorola developed a lead free solder dras. Not only will this product help it comply with the EU initiatives; Motorola now gets a percentage of the profits its suppliers earn from sales of the product (Rock and Angel 2005).

Lessons for Developing Countries

What can developing countries in Latin America and elsewhere learn from Asia's IT experience? After all, Taiwan rose to prominence in an earlier era, before the advent of such intense global competition and the constraints on targeted industry policy now imposed by the World Trade Organization. India and China derive enormous benefits from their huge internal markets. Even if they were willing, few developing countries have the kind of leverage China has in bargaining with MNCs for technology and other benefits.

Despite some unique characteristics, the experience of Asia offers four lessons for Mexico and other developing countries seeking to upgrade their IT or other industries to be globally competitive.

First, governments must take a proactive role in building institutions and enacting policies that attract foreign investment and nurture local capacities for production and innovation. Especially important is to improve *learning capabilities*. Without strong capacities to learn, local firms are unable to reap spill-overs from FDI that promote industrial upgrading. But the cultivation of learning capacities is outside the purview of MNCs. "The Malaysian experience . . . shows that MNCs are unlikely to invest in building local capabilities and institutions when they cannot appropriate the returns fully and when they do not have effective control over their coordination. There is thus a need for significant government intervention to overcome market failures and to resolve collective action problems." (Rasiah 2003, p. 330)

Government intervention took two forms in Asia. One form consisted of general, economy-wide policies that were aimed at improving overall national capacities for production and innovation, such as science and technology policies and investment in education (primary, secondary and tertiary), vocational training, and infrastructure, including transport, waste management and communication. The other form consisted of industry policies that targeted the IT sector with credit subsidies, government-industry R&D partnerships, and training programs.

In a major study of the development of the high-tech industry in Korea, Brazil, and India from the 1970s to the early 1990s, Peter Evans (1995) suggests that governments combined general and targeted policies in four distinctive roles:

Custodian   The state regulated markets in ways that favor targeted sectors.

Demiurge   The state served as the investor, risk taker, research and development hub, and information finder to seek out areas where the private sector would not initially invest.

Midwife   The state worked to spawn private sector start-ups and spin-offs.

Husband   The state actively nurtured the capacities of targeted domestic private firms through education, training, credit subsidies and partnering in research and development.

Evans observed that the most successful countries played strong and effective roles of midwifery and husbandry. Brazil and India played strong custodian and demiurge roles and fell prey to rent seeking. South Korea stood out as the nation that was able to help establish national firms through midwifery and through husbandry ushered them to the point that they stand on their own feet.

The experience of Taiwan and later, India and China, confirm the importance of midwife and husbandry policies, especially investment in human capital, including general and technical education and training. In his study of the IT sector in Korea, Lim (2003) also emphasizes the role of R&D and suggests that it is important not to worry in early stages about traditional indicators of performance, such as the number of patents. What is most important is to develop a substantial number of experienced researchers capable of playing a role in R&D in the private sector.

The second lesson from the East Asian experience is that it is important to promote domestic diffusion of computers, both to modernize the economy and to create demand for local firms. A focus on production for export alone is unlikely to generate the kind of spill-overs that can substantially boost local firms and keep MNCs from becoming footloose. Moreover, the larger the domestic market, the more bargaining power the government has in negotiating favorable terms from MNCs.

The third lesson is that gaining entry into the global IT industry is difficult, not only because of the oligopolistic structure and competitive dynamics of the industry itself but also because the field is already very crowded. Developing countries looking for industries in which they can develop substantial global markets may want to look elsewhere. Indeed, even in Latin America, Amsden suggests that countries look for new opportunities to other high-tech industries, such as petro-chemicals, pharmaceuticals and bio-tech (2004, p. 87). As we will see in the case of Mexico, leveraging FDI through joint ventures is essential.

The fourth lesson is that without the proper policies in place, IT growth can come at a significant cost to human health and the environment. Moreover, neglect of environmental regulation can reduce future export competitiveness as the EU and other OECD countries raise environmental standards.

5

Mexico's Bid for a Place in the Global IT Industry

With strong local capacities for electronics manufacturing, Mexico seemed poised to ride the coattails of the global IT boom of the 1990s. To the Mexican government, the path to building globally sustainable manufacturing industries, including IT, lay in complete and uncompromising liberalization. The strategy was to attract MNCs en masse to produce for export, enabling Mexico not only to become a manufacturing hub but to gain the skills and technological sophistication to move up the IT value chain and to drive broad-based economic growth.

By some indicators, the gambit worked remarkably well. Being next-door to the world's largest and fastest-growing IT market gave Mexico a unique, location-specific asset. Coupled with low wages and the industry's shift toward outsourcing (see chapter 3), geographical proximity to the United States made Mexico very attractive to IT flagships and CMs in the 1990s. By the end of the decade, Mexico seemed to be emerging as the industry's manufacturing cluster for the North American market, replacing East Asia and southern US states.

In an abrupt reversal, however, the flagships started scaling back in Mexico in 2001, shifting production to China and other Asian countries and leaving the future of Mexico's IT industry in limbo. While its geographical assets remained intact, Mexico's supplier capacities were damaged, rather than enhanced, by the FDI wave. Moreover, wages in Mexico became increasingly high compared to East Asia. By 2006, the industry was far from dead—a substantial amount of assembly and sub-assembly work for a variety of electronic product lines continued apace. However, the dynamism that infused hopes of an IT-led economic boom had evaporated.

Without a significant change either in market conditions or government policies—or both—it is unlikely that Mexico will recapture it.

From Big Government to "Big Bang" Liberalization

Mexican endogenous capacities for IT manufacturing were seeded and cultivated by import-substituting industrialization (ISI) policies from the 1940s to the 1980s. In the 1940s, Mexican firms began manufacturing radios and radio components, adding televisions and related parts in the 1950s. In their heyday, Mexican firms accounted for 85–95 percent of the value-added content of television production. Mexican firms such as Majestic co-located outside of Mexico City alongside multinationals such as Philips, General Electric, and Magnavox and supplied a growing domestic market (Lowe and Kenney 1999). IBM set up operations in Mexico City in 1957 and began producing manual typewriters (Palacios 2001).

In the 1970s, in line with the ISI strategy, the government targeted computer industries as part of Mexico's National Council on Science and Technology's (CONACYT) strategy to increase Mexico's national self-sufficiency in technology. CONACYT set forth the "Program to Promote the Manufacture of Electronic computing Systems, their Central Processing Units, and their Peripheral Equipment," which became commonly known as the PC (Programma de Computadoras). The program's stated goal was to develop a domestic computer industry (supported by the surrounding electronics industry) that could not only serve the domestic market but also emerge as an important export industry. Specifically, the government target was that domestic supply would cover at least 70 percent of domestic demand by 1970 (Peres 1990). According to Dedrick, Kraemer, and Palacio (2001), the program had six components:

Market access   Access to the Mexico's computer market was strictly limited to firms that would meet the goals of the program.

Foreign investment   Foreign firms could own no more than 49 percent of domestic firms.

R&D   Foreign firms had to invest 3–6 percent of gross sales in R&D as well as create research centers and training programs.

Government procurement   Firms registered in the program received preferential procurement benefits.

Domestic content   For personal computers, 45 percent of the value added had to be domestic parts and components. For mini-computers, it was 35 percent.

Fiscal and credit incentives   New Mexican-owned firms were eligible to receive fiscal credits and low interest loans from government development banks.

Mexico's program was an extraordinary success, especially in view of the 1982 debt crisis. By 1986, Mexico had succeeded in attracting major international firms (at only 49 percent ownership) and had built a domestic industry. According to a comprehensive assessment of the program (Peres 1990), 60 foreign and domestic companies had registered in the program by 1987, employing 6,000 workers. The value of total production amounted to approximately $400 million, with micro-computers accounting for half, mini-computers for 37 percent, and peripheral equipment for twelve percent. By 1987, about half of all production was exported to markets in the United States and Canada.

In search of both Mexican domestic markets and export platforms, the foreign firms that came were IBM, Hewlett-Packard, Digital, NCR, Tandem, and Wang. IBM and Hewlett-Packard accounted for 63 percent of all Mexican computer production. Other foreign firms accounted for approximately 18 percent, and fully owned Mexican firms for another 18 percent.

The leading Mexican maker of personal computers during the ISI period was Printaform. By 1986, Printaform product quality was considered to be comparable to that of IBM and Hewlett-Packard—triggering UNAM (Mexico's largest university) to purchase Printaforms over the MNC competition. After liberalization, Printaform's share of the PC virtually evaporated, although the firm continues to produce office equipment. Lanix, another firm spawned during the ISI period, was spun off from a micro-computer firm and today sells personal computers in Mexico and Latin America (Dedrick et al. 2001).

Mexico's domestic computer sales also grew rapidly under the PC program. By 1986, Mexico's domestic computer market was second only to Brazil's in Latin America (Whiting 1991). Remarkably, national supply was 56 percent of domestic demand in 1987, only 14 percentage points short of CONACYT's goal of having 70 percent. Joint ventures supplied half of Mexican demand, and the other half was met by wholly Mexican-owned firms.

The Defection of IBM

Despite its success, the Programa de Computadoras was dealt a fatal blow in 1985 with the defection of IBM. Involved in the planning stage for its first computer manufacturing plant in Guadalajara, IBM insisted that it be exempted from compliance with the 49 percent rule and be allowed whole ownership of the plant. IBM had several rationales. First, the company was eager to expand its marketing efforts to both Mexico and Latin America, having focused almost solely on penetrating the US computer market. The greater control that would come with whole ownership would facilitate IBM's strategic marketing. Second, the company found Guadalajara to be attractive as a computer manufacturing site because of its network of local suppliers. Third, IBM wanted Mexico to be seen as an alternative to the state-led computer policies being advocated in countries like Brazil and India. According to Whiting (1992, p. 205), "IBM hoped its Mexican project would demonstrate the benefits to underdeveloped countries of allowing a less nationally controlled computer industry."

In responding to IBM's demand, Mexico was between a rock and a hard place. IBM had first come to Mexico in 1957 to produce sewing machines and was the first mover and the strongest firm in the global computer industry. On the one hand, the Mexican government generally opposed IBM's demand, though it had to take it seriously because of IBM's global market share. On the other hand, the government saw the exemption as a way to muscle in on IBM's dynamic personal computer market. If Mexico's newly spawned firms could serve as suppliers, IBM's expanding markets would propel the growth of Mexican firms.

Mexico had significant leverage—its large and growing market, its proximity to other Latin American markets, and its proximity to the United States. Mexico brokered a deal whereby IBM was granted the exemption and given whole ownership of its new plant in exchange for an

$11 million Center for Semiconductor Technology. Based in Guadalajara, the Center was to train Mexican engineers and developers over the years 1988–1994.

The granting of the exemption was a watershed event. Soon after, Hewlett-Packard demanded and received an exemption from the joint venture rule—without any conditions or requisite commitments toward local development. By 1987, other foreign firms were also allowed to work outside of the PC program. The IBM exemption laid the groundwork for full-blown liberalization.

Best Policy: No Policy

In 1990, the Mexican government began dismantling the PC program as part of its "big bang" rapid liberalization process which culminated with the signing of NAFTA in 1994. Three factors contributed to Mexico's change in vision and policy for the IT sector. First, the debt crises of the 1980s created a need for foreign exchange. The Mexican government was keen to increase FDI inflows and boost exports, including in the IT sector. Second, the global IT industry was restructuring, creating both new opportunities and new pressures for developing countries to abandon inwardly focused growth and to integrate into MNC global supply chains. Third, politics in Mexico were shifting toward free-market ideology. In a radical break from the past, President Salinas presented an economic vision that relied solely on market forces to allocate resources and generate growth. The computer sector was an early target for the Salinas approach, described by an administration official as "the best policy is no policy."

While Salinas began liberalizing the computer sector in 1990, it was during the NAFTA negotiations that the old national government-assisted PC program was demolished. Under NAFTA, foreign investment was fully liberalized, tariffs fell to zero, and policies favoring domestic producers, including government procurement, local content requirements, and R&D requirements were abolished. Interestingly, some of these industry policies were transferred to NAFTA's regional "rule of origin" section, driving many US and Asian firms to locate to Mexico (CEPAL 2000).

The Zedillo administration which succeeded Salinas continued to put its faith in market forces to drive IT development. However, Zedillo envisioned that the government would complement market forces by promoting IT use throughout the country, improving the basic telecommunications infrastructure, and increasing R&D to develop local firms that could supply niche markets. The plan never made headway, however, because no funds were allocated for its administration and the institutions charged with carrying it out had little to no coordination. The Fox administration that succeeded Zedillo also acknowledged the need for industry support but argued that it simply did not have the funds and capacity to carry it out (Dussel et al. 2003).

Electronics Triumph

The laissez-faire strategy was an early brilliant success—at least in terms of attracting investment and increasing exports. Between 1994 and 2000, foreign direct investment in the electronics sector quintupled and the value of exports quadrupled. At their peaks, exports from Mexico's electronics sector totaled $46 billion in 2000, and FDI inflows totaled $1.5 billion in 1999.

By 2000, electronics was a key component of the Mexican manufacturing economy, accounting for nearly 6 percent of manufacturing output, 27 percent of all exports, 9 percent of employment, and 10 percent of FDI. Electronics are Mexico's largest manufactures export, and are second only to autos in terms of manufacturing GDP and employment (Secretaria de Economia 2003).

Nationally, the electronics industry in Mexico is spread across three geographical regions: the border region, Mexico City and environs, and the western region, especially the state of Jalisco (figure 5.1). Production in the US-Mexico border region is almost exclusively dedicated to audio and visual products for re-export to the United States. Sanyo, Sony, and Mitsubishi account for the bulk of FDI there. Another cluster of electronics firms is centered around Mexico City, home to nearly one-fourth of Mexico's population and accounting for the bulk of domestic demand for electronic products. Most firms in the Mexico City region produce home appliances and consumer electronics for the Mexican market.

The western state of Jalisco is home to Mexico's IT sector. Fueled by large FDI inflows, Mexico's IT industry became increasingly competitive during the second half of the 1990s. Mexico's share of world IT exports

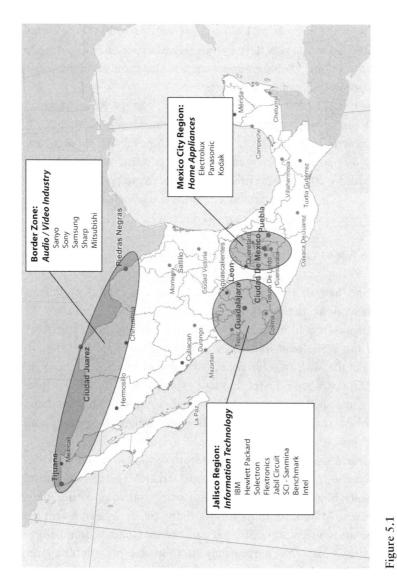

Figure 5.1
Electronics clusters in Mexico. Source: Global Development and Environmental Institute. Information from CADELEC 2004.

ballooned from 0.8 percent in 1985 to 3 percent in 2000 (Dussel et al. 2003). By 2001, Mexico was the eleventh-largest exporter of IT products in the world economy. Among developing countries, only Malaysia and China outpaced Mexico in global IT exports.

However, rapid MNC-led growth came at the expense of Mexico's domestic IT firms, which were virtually wiped out. Moreover, after 2001, MNC-led growth itself came to an abrupt halt. Since 2000, exports have dropped by 60 percent, FDI has fallen by 123 percent, and 60,000 jobs have been lost. By 2003, most IT flagships had relocated productive functions to China. The next two sections describe the golden age and the era of flattened expectations for Mexico's IT industry.

The Golden Age of Guadalajara's IT Industry

In the second half of the 1990s, Guadalajara, the capital of Jalisco, rapidly emerged as the heart of the Mexico's MNC-dominated IT industry. "High-Tech Jobs Transfer to Mexico with Surprising Speed," proclaimed the *Wall Street Journal* in an article dated April 9, 1998. To the surprise of the *Wall Street Journal*, Guadalajara had become home to a booming IT assembly industry that had exported $7.7 billion worth of IT products in that year. It was a short-lived triumph. Two years later, the cover of *MexicoNow*, a leading Mexican business magazine, read "Rescuing Mexico's Electronics Industry."

Guadalajara was founded in the sixteenth century and, with a population of 4 million, is now Mexico's second-largest city. It is a beautiful city, with leafy bouvelards, a large *zocalo* or central square, museums, and universities. Millions of tourists flock to the city each year to visit the sites and see the numerous murals by the renowned Guadalajaran painter Jose Clemente Orozco. Not surprisingly, local inhabitants exhibit a large measure of civic pride.

Guadalajara has a 40-year history of electronics manufacturing. The first electronics firms came to Guadalajara as a result of Mexico's maquiladora program to draw FDI to areas away from the US-Mexico border. Motorola and Burroughs opened plants in Guadalajara in 1968, followed by IBM in the 1970s and by Wang, Tandem, and Kodak in the 1980s. In the 1970s and the 1980s, numerous local electronics firms

Table 5.1
Major multinational corporations in Jalisco's IT sector. An asterisk indicates that the firm had shut down or moved as of 2004. PCB stands for printed circuit board.

	Ownership	Activity
Flagships		
Hewlett-Packard	US	Marketing, logistics, procurement, call center, sales, marketing
IBM	US	Telecommunications manufacturing
NEC*	Japan	Telecommunications manufacturing
Motorola*	US	Telecommunications manufacturing
Lucent Technologies*	US	Manufacture of compact discs and CDs
Kodak	US	Testing and design
Intel	US	"
Contract manufacturers		
Flextronics	US	PCB, game boxes, PCs, laptops, peripherals
Solectron	US	"
SCI-Sanmina	US	"
Jabil Circuits	US	"
Yamaver	Japan, Belgium	PCB assembly
NatSteel	Singapore	"
Pemstar*	US	"
KRS International*	Germany, Belgium	"

linked to the MNCs emerged as a result of the ISI policies. Mexican firms such as Zonda, Kitron, Microtron, Electrónica Pantera, and Wind either spun off from the MNCs to produce their own PCs or served as suppliers to the MNCs (Wilson 1992; Palacios 2001).

During the boom years of the 1990s, a new influx of global, mostly US- headquartered, flagships expanded or newly came to Guadalajara, including Hewlett-Packard, IBM, Intel, and Lucent Technologies. The Japanese flagship NEC also located in Guadalajara (Woo 2001).

In the mid 1990s, the global IT industry was well into the process of reorganizing into a global production network model, with manufacturing functions largely outsourced to independent firms. Bypassing the possibility of outsourcing to Mexican firms, the flagship companies recruited

Table 5.2
High-tech project losses in Jalisco, 2001–2003. Source: Dussel 2005.

	Investment (US millions)	Employment	Destination
Hard disk drives	108	4,250	China
Components	30	1,200	China
Communication systems	n.a.	3,720	China
Cell phones	n.a.	400	China
	24	1,493	China
	25	1,095	China
Semiconductors	200	2,100	Philippines
Electronic cards	24	1,049	China
	70	925	Malaysia
Printers	12	1,900	China
Printing systems	3	295	China
Electronic equipment	3	300	China
Telecommunications	15	2,500	China
Total	514	21,227	

the large CMs in which they were consolidating global manufacturing. The firms recruited were mostly large US-based contract manufacturers—SCI-Sanmina, Solectron, Jabil Circuit, Flextronics, and Pemstar. Non-American CMs that came to Guadalajara were Natsteel (Singapore), Yamaver (Japan and Belgium), and KRS International (Germany-Belgium). All arrived in 1996 and 1997 (Palacios 2001).

Why did the global IT flagships choose Mexico in general and Guadalajara in particular as a place to build a manufacturing base in the 1990s? Both offer a number of location-specific assets, many stemming from the fact that Mexico shares a relatively peaceful and, after NAFTA, economically open border with the world's largest and fastest-growing market for high-tech products. Location-specific assets include the following:

Speed to market Shipments from Guadalajara to California's port of Long Beach take less than a day. Shipments from China, Japan, and Malaysia take 15, 12, and 23 days, respectively (CANIETI 2003).

The new rules of origin under NAFTA Printed and electronic circuits and motherboards must be manufactured in North America (Alba 1999).

The new tariff structures under NAFTA Mexican tariffs for IT imports exceeded 20 percent in the 1980s under the PC program but were lowered to zero under the NAFTA (Dedrick et al. 2001). For non-NAFTA countries the tariff on the IT sector remains 20 percent. In addition, for Mexico the United States has decreased its import tariff on IT to zero (Dussel 2003).

The PITEX Program The PITEX program allowed firms to import their inputs duty-free as long as more than 65 percent of their product was exported (Dussel 2003).

Jalisco's Economic Promotion Law The law temporarily reduces or eliminates state and municipal taxes for foreign investors (Alba 1999).

Technological sophistication Guadalajara has five major universities and numerous technical schools and industrial parks that can host research activity and generate an adequately skilled workforce (Palacios 2001).

Weak labor unions and low wages Unions were seen as "easy to handle and even hands off." Wages in Guadalajara's IT sector were one-seventh those in the United States (Palacios 2001; CANIETI 2003).

Infrastructure The Guadalajara area has a good supply of water.

Lifestyle Guadalajara offers comfort, recreation, and culture (Palacios 2001).

All these factors came together after 1994. NAFTA's rules of origin, tariff preferences, and Mexico's wage and union structure made Mexico the obvious choice for investment, Guadalajara won because of its pre-NAFTA manufacturing experience in the sector, its speed to market and universities, and complimentary national (PITEX), state, and local incentives. Yet geographical proximity offering speed to the US market has to be underscored.

   Government policy also deserves note. The state government of Jalisco government supplements federal programs with regional incentives and FDI promotion programs to attract MNCs, such as the Economic Promotion Law listed above that reduces or eliminates state and municipal taxes for firms that located to the region. The Guadlalajara branch of the national chamber of commerce for the IT industry, CANIETI (Camara Nacional de la Industria, Electronica, de Telecomunicaciones e Informatica) works to attract large MNCs to the region and puts on numerous trade shows and workshops.

The Jalisco government also helped to establish a regional organization to help build a local supplier base. CADELEC (Cadena Productiva de la Electronica) was founded in 1998 with funding from CANIETI, the United Nations Development Program, and two other federal agencies. CADELEC's mission is to match locally based suppliers (foreign or domestic) with the large MNCs. They also compile data on the IT sector in Jalisco and globally, serve as a resource and promotion organization for foreign investment, and host conferences and workshops (CADELEC 2004; Palacios 2001).

Buoyed by its location-specific assets, foreign investment surged into Guadalajara, generating a boom in both exports and employment. Foreign investment tripled between 1995 and 1998 (but tapered off by 2002 to its 1995 level). The United States was the main source, accounting for 91 percent of total investment in Jalisco's IT sector between 1990 and 2002. Singapore and Taiwan accounted for 7 percent and the rest came from Japan, Germany, and Mexico (SEPROE 2001). Domestic investment amounted to only 1 to 4 percent of total investment per year over the entire period (Dussel, Lara et al. 2003).

IT exports quadrupled between 1995 and 2002 and by 2002, accounted for more than 61 percent of all exports from Jalisco. The state's second-largest export was paper and cardboard at 13.5 percent (CADELEC 2004). Indeed, in 1999, IT exports from Jalisco totaling $9.3 billion surpassed the value of Mexico's entire crude oil production, which amounted to $8.9 billion (Alba 1999).

The bulk of Jalisco's IT exports go to the United States, with a small amount going to the rest of Latin America, Europe, and Japan. Between 1990 and 2001, Mexico's IT exports to the United States grew by nearly 27 percent per year.

IT growth has had significant impacts on the state and regional economy. In value-added terms, the IT sector grew by a factor of 5 between 1993 and the boom's peak in 1998—compared to a doubling for manufacturing as a whole. Employment in IT grew by 131 percent during the period, compared to 61 percent for the state as a whole. No other sector in Jalisco saw such growth. Indeed, most others shrank. IT became a trade powerhouse for the state and region as well. By 1998 exports from the IT sector accounted for nearly 84 percent of all exports from Jalisco (Maquilaportal 2005).

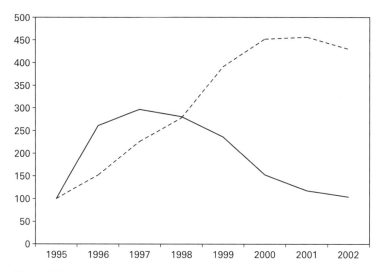

Figure 5.2
IT exports and foreign investment in Jalisco, with 1990 = 100. --: exports. — foreign direct investment. Source: author's calculations, based on CADELEC 2004.

From Bienvenidos to Adios

A combination of foreign and domestic factors caused tough times for the IT sector in Guadalajara. Between 2000 and 2004 there were major job losses, and many firms moved production functions overseas, mainly to China. While Jalisco maintained a place in the global IT commodity chain, it is a relatively stagnant place dependent on speed to the US market and relatively low wages. Industry analysts, economists, and even government policymakers have largely abandoned hope that Mexico will become a "silicon valley," gaining skills and technological sophistication to enable to move up the value chain. Even its position as an assembly and sub-assembly site is vulnerable to rapidly changing global market conditions and the cutthroat competition that drives the industry.

Why did Mexico fail to capitalize on the post-NAFTA investment wave to capture a secure and dynamic place in the global IT industry? The largest part of the blame lies in Mexico, with failures of both commission and omission in government policy. But much is due to larger changes, both in the global economy and the global IT industry. Indeed, the main failures of policy stemmed from the failure to understand and shape appropriate responses to global economy and industry dynamics.

In the 1990s, investment occurred at a feverish pace not only in Guadalajara and North America but throughout the global IT sector, generating excess capacity. When the high-tech financial bubble burst in 2001, demand for final products fell abruptly and the industry went into a spasm of restructuring. Some Guadalajara-based firms downsized dramatically, especially companies like Lucent who were heavily over-capitalized. All scrambled to reduce production costs.

While all IT firms were affected, US firms were hit the hardest. As late as 2004, US firms were still afflicted with excess capacity and inventory that had not entered the market. The continuing slow growth in the US market was strongly felt in Mexico: 64 percent of Guadalajara's IT exports go to the United States.

There were also significant developments in the global economy due to increasing economic globalization. By 2001, Mexico was not as attractive as a low-wage platform as it had been 6 years earlier. Mexico's wages are now the most expensive of those in the global electronics production network. In 2001, Mexico's hourly wages averaged $2.96, compared to $2.58 in Hungary, $2.17 in Malaysia, and 72 cents in China. By 2005, Mexican hourly wages had fallen slightly, to $2.40. However, new competitors had emerged: the hourly wage was $2.15 in Taiwan and $1.10 in India. Though the average hourly wage in China had increased to 95 cents, the gap between Mexico and China was still yawning (Sherman 2004).

A major global development was the 2001 accession of China to the WTO—virtually at the same time that the slump hit the IT industry. China's economy has been growing at a rate of more than 8 percent per annum since 1978. China has built both substantial manufacturing capacities and domestic markets, including in IT products. China's embrace of WTO rules made it easier to manufacture and export products in China. Gone are high tariffs for infant industries and local content requirements requiring foreign firms to buy or train locals. On the other hand, subsidies to foreign firms remain in place, including cheap land and infrastructure and conditioned access to China's domestic markets.

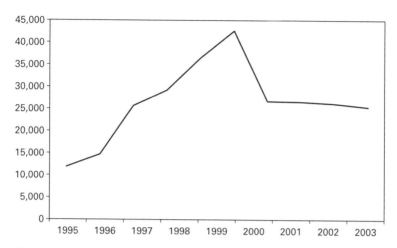

Figure 5.3
IT employment in Jalisco, 1995–2003. Source: CADELEC 2004.

The combination of industry shake-out and the new attractiveness of China for IT manufacturing hit the global flagships hard in Guadalajara. Lucent suffered the worst losses, with a complete shutdown of its state-of-the-art Jalisco plant—designed for 12,000 workers—and a shift in all operations to Brazil and China. NEC and Motorola also shut down their Guadalajara operations. By 2003, only two major IT flagships still operations in Guadalajara: Hewlett-Packard and IBM. Though both were originally engaged in manufacturing, neither has a trace of that past.

Hewlett-Packard had shifted from manufacturing to research and development in the late 1990s. However, in 2003 it shut down its R&D operations and moved them to Boise, Idaho. As of 2006, HP Guadalajara has carved out a niche as a Spanish-language call center called Jaliscience for sales and customer support. Hewlett-Packard customers who press 2 for Español are connected to this call center, which is called Jaliscience. In addition, Hewlett-Packard coordinates logistics for all its Latin America offices out of Guadalajara. IBM took longer to switch out of manufacturing in Guadalajara (it never moved into R&D) and, like Hewlett-Packard, had become by 2003 completely service oriented, focusing on marketing and software development.

The CMs in Guadalajara have been hard hit as well. According to interviews we conducted in the period 2003–2005, more than 12,000 workers were laid off by CMs: nearly 7,000 by Flextronics, 2,000 by Jabil, and 3,500 by SCI-Sanmina. The trade newsletter *Maquila Portal* reported in 2004 that "the recent economic slowdown and the exodus of companies to China caused Solectron to fire 7,000 employees, which has left a difficult row to hoe," and that "the company is currently function- ing with 70 percent of its original workforce and probably won't open a new production plant in Mexico for another two years." Pemstar shut its Guadalajara plant in 2005.

For the most part, CMs have been able to redirect their generic manufacturing capacities to other products and clients. Jabil Circuit, for instance, shifted production to communications switches, specialized hand-held credit card processing machines, Internet firewalls, and elec- tronic controls for washing machines (Luhnow 2004). Solectron is assem- bling components for mainframes and AX-400 conductivity transmitters. SCI-Sanmina now assembles MRI scanners for Phillips and electronics auto components for Ford and GM (Contreras 2006). Not all of the CMs were able to upgrade with this agility, and national firms remained out in the cold. Indeed, most of the CMs have resorted to Internet-based open supplier bidding.

According to Dussel (2005), nearly 22,000 workers lost their jobs in Jalisco's IT industry—including flagships, contract manufacturers, and local supplier firms—between 2001 and 2003. Over the same period, investment shrank by $514 million. The contraction affected a wide range of IT products, including hard-disk drives, electronic cards, and semiconductors, as well printers and cell phones.

Where did the jobs and investment go? In terms of numbers, most high-tech investment projects shifted to China (Dussel 2005). However, the Philippines and Malaysia captured two large projects accounting for more than half of investment and nearly 3,000 jobs.

With its low wages, strong networks of local suppliers, and strategic government support, China is Mexico's main global competitor. In United States, EU and Japanese IT markets as a whole, Mexico lost market share between 2000 and 2004 in each of three key product lines—comput- ers, peripherals, and telecommunications—while China gained hugely. China is even challenging Mexico's in the US market. China's IT exports

Table 5.3
Percentage of imports from developed countries, including EU (15), Japan, and US. Source: United Nations 2006.

	Computers	Peripherals	Telecommunications
China			
2000 market share	7.53	8.52	8.05
2004 market share	27.97	22.38	18.65
Percentage point change	20.44	13.86	10.60
Mexico			
2000 market share	4.41	3.00	7.76
2004 market share	3.86	1.70	6.14
Percentage point change	−0.55	−1.29	−1.62

to the United States grew at 60 per year since 1990 and accounted for 12 percent of the US market—just one percentage point below Mexico (UNDP 2001).

China out-competes Mexico in nearly every aspect of production costs, including wages, costs of raw materials, construction costs, and cost of capital (interest rate). The only area where Mexico outcompetes China is in transportation costs. Mexico also has a slight advantage in income tax. But these cost advantages pale beside a wage differential of nearly 600 percent and a gap in construction costs of 400 percent. Specific to the electronics industry China outcompetes Mexico in the US market despite the fact that China's tariff is 6 times Mexico's (Dussel 2005).

Despite the losses after 2001 and the continuing cost disadvantages, Guadalajara is far from an IT ghost town. Mexico will always retain the costs of transport and speed-to-market advantages of geographic proximity. Moreover, the advantage of lower transport costs will be enhanced if—as seems likely—oil prices continue to rise. SCI-Sanmina is very conscious of speed as a marketing strategy and promises US customers that it will meet any order from the United States within 48 hours. Marco Gonzalez Hagelsieb, senior vice-president of the company's Mexico operations, puts it this way: "If you were to order ice cream from China you would get five containers of vanilla. Whereas Mexico is a Baskin-Robbins: we can mix and match flavors and deliver the ice cream the next day." (Contreras 2006)

As of mid 2006, what is left in Jalisco are just two flagship IT firms with service operations, a handful of US-heaquartered contract manufacturers scrambling for orders, and a small clutch of struggling local suppliers. Acknowledging the shortcomings of past policy, the Fox administration shifted its goals to focus on IT goods that are "created in Mexico" rather than "made in Mexico."

Is such an approach too little too late? The Mexican IT industry is highly vulnerable to US market conditions, global shifts in industry governance, and other global conditions beyond its control. If oil prices rise dramatically, Mexico's geographic advantage, which enables low transport costs, could override its relatively high wages. If and when US demand increases, Mexico could receive a surge of investment, with corresponding growth in production, employment and exports. Conversely, when the United States is in the middle of a slowdown, the IT industry in Mexico will feel the pain because it lacks sources of demand beyond the United States. A US official put it this way: "They're still sending 90 percent of their products to the US, and as soon as our economy goes south, so will theirs." (Contreras 2006, p. 37)

6

Silicon Dreams, Mexican Reality

The golden age of the Mexican IT industry lasted for less than a decade before the global flagship companies redirected manufacturing to East Asia. What did they leave behind? During their tenure in Guadalajara, did they and the contract manufacturers they brought with them generate knowledge spill-overs that could help local firms move up the IT value chain and spark broad-based regional development? What did national and state governments do to help capture spill-overs?

When the golden age dawned, Mexican firms had substantial capacities to manufacture IT products and components. Indeed, two influential studies produced in the 1980s—Patricia Wilson's *Exports and Local Development* (1992) and Harley Shaiken's *Mexico in the Global Economy* (1990)—saw great promise for the emergence of a globally competitive IT industrial cluster in Guadalajara with strong local linkages. While they found only limited evidence of FDI spill-overs, both studies concluded that the sector was poised to flourish because local suppliers were abundant. According to Wilson (1992, p. 120), it was just a matter of the government playing a nurturing role: "The Mexican government can increase the local linkages not only of the locally owned maquiladoras but also of some of the foreign-owned maquiladoras. In Guadalajara, the sector of locally owned producers and the sector of foreign-owned electronics producers show particular potential."

In the event, the Mexican government eschewed proactive strategic industrial policies aimed at increasing either the propensity of multinational corporations to supply or the capacity of local firms to absorb spill-overs. Moreover, lack of access to local credit and Mexico's macroeconomic policies created an economic environment that favored foreign over domestic firms. Rather than flourish, local IT manufacturing and supplier firms were all but obliterated.

This chapter spotlights the interaction between foreign firms—global flagships and CMs—and local firms in Guadalajara during the golden age when Mexico seemed poised to be emerging as "Silicon Valley South."

Backward Linkages—Local Manufacturers and Suppliers

The Mexican government's liberalization strategy aimed to attract global IT flagship firms to locate in Guadalajara, bringing with them their core manufacturing capacities while generating knowledge spill-overs to local firms primarily through backward supply linkages. The hope was that spill-overs would facilitate innovation and growth by local IT firms, stimulating the entire regional economy.

Knowledge spill-overs from MNCs may flow through several channels to local manufacturers and suppliers. Increased demand may allow local suppliers to gain economies of scale, thus increasing efficiency and/or revenues that can be used to invest in new technology. Higher profitability of local firms can also crowd in domestic investment, spurring expansion of existing firms or the starting up of new ones.

Spill-overs can also flow to local contractors and suppliers through MNC requirements to upgrade technology and/or to adopt higher and more consistent quality standards. MNCs may provide training to local firms to do so. The potential for positive spill-overs is further enhanced if MNCs partner with domestic universities or firms in joint R&D projects that can lead to upgrading along the IT value chain over the medium to long term. Negative spill-overs occur through backward linkages if the presence of foreign firms reduces the efficiency or the competitiveness of domestic supplier firms or drives them out of business altogether. They also occur if FDI crowds out domestic investment.

Like the global IT industry in general (chapter 3), Guadalajara-based MNCs began in the mid 1990s to outsource their manufacturing functions to third-party firms, potentially presenting a golden opportunity for the Guadalajara region. Indeed, in the early 1990s nearly 50 local firms were involved in some kind of electronics manufacturing—a number of whom were already CM assemblers (Wilson 1992). This is exactly the opportunity that Taiwanese firms had seized some years earlier (chapter 4). Existing CMs were able to scale up their activities and directly contract with the leading tier flagship firms, thus capturing the market and opportunity to serve as CMs (Amsden and Chu 2003).

By the late 1980s, the IT sector was producing about $400 million worth of products per year and employing 6,000 workers. Working alongside global flagships such as IBM, Hewlett-Packard, Digital, Kodak, NCR, Tandem, and Wang were domestic computer manufacturers such as Scale and Electron Computers and wholly owned or joint-venture assembly manufacturers such as Unisys, Cumex, and Mexel. The bigger firms were supplied by numerous local niche firms, including Electrónica Pantera and Microtron (Wilson 1992; Palacios 2001; Rivera Vargas 2002).

When the trend toward outsourcing began, a number of Mexican firms seemed poised to grab the opportunity. A major contender was Unisys, a 100 percent Mexican-owned CM of computers and peripherals. Unisys was a spin-off from the US-based firm Burroughs. At its peak, Unisys employed one-seventh of the entire Guadalajara electronics workforce. Wilson argues that Unisys, as a nationally owned firm, "did more than any of the foreign electronics maquiladoras in Guadalajara to develop a domestic supplier base." Eighteen percent of its inputs were supplied by firms located in Guadalajara or elsewhere in Mexico. Among these locally sourced inputs were terminals and connectors, metal parts, and cables. Furthermore, Unisys spawned two spin-off firms: Compubur and Electrónica Pantera (ibid.).

Rather than turn to local firms, however, the global flagships recruited US-based CMs to set up shop in Guadalajara. By the mid 1990s, the US-based CM giants Jabil Circuit, SCI-Sanmina, Flextronics, and Solectron (along with NatSteel from Singapore) had established plants in Guadalajara that conducted virtually all of the manufacturing for Hewlett-Packard and IBM and the other flagships. The flagships' strategy in Mexico was part of a larger trend to consolidate supply relationships in a few global suppliers, rather than to contract with a multiplicity of CMs to supply national or regional markets (chapter 3).

The global IT industry's shift toward consolidation of outsourcing in a few CMs occurred at exactly the same time that Mexico was shifting high-tech strategies and implementing its neo-liberal FDI strategy. The result was to remove any incentives to global flagships to outsource manufacturing to Mexican CMs. Indeed, Mexican trade and exchange rate policies generated a positive bias both toward foreign manufacturing firms and imported inputs (see below).

The result of the changes in global industry strategy and Mexican policy was a virtual wipe-out of the Mexican IT industry, including manufacturers, assemblers, and suppliers. Rivera Vargas (2002) found that there was a 71 percent decline in the number of domestic firms between 1985 and 1997 (table 6.1). In our interviews, we learned that 13 of the 25 domestic firms still in existence at the end of 1997 had gone out of business by 2004.

Table 6.1
Closings of IT plants wholly or partly owned by Mexican firms. Source: Woo 2001; Rivera Vargas 2002; interviews.

Firm	Ownership (%)	Activity
Cumex Electronics	50/50 Mexico-US	Contract manufacturer of printed circuit boards
Mitel	51/49 Mexico-Canada	Telephone components
Phoenix International	50/50 Mexico-US	Plastic injection
Encitel	100 Mexico	Contract manufacturer of printed circuit boards
Info Spacio	100 Mexico	Contract manufacturer of printers
Logix Computers	100 Mexico	Design and manufacture of personal computers
Mexel	100 Mexico	Contract manufacturer of printed circuit boards
Unisys	100 Mexico	Contract manufacturer of computers and peripherals
Electron	100 Mexico	Design and manufacture of personal computers
Scale Computers	100 Mexico	Design and manufacture of personal computers
Advanced Electronics	100 Mexico	Design and manufacture of printed circuit boards
Compuworld	100 Mexico	Contract manufacturer of hard drives
Microtron	100 Mexico	Buffers and carton packages

The Wipe-Out of Local Suppliers

The wipe-out occurred for two reasons. On the one hand, the global flagships contracted with foreign CMs for manufacturing and assembly. On the other hand, the CMs turned toward foreign rather than local suppliers for parts, components and even services.

Mexican CMs like Unisys had relied on local suppliers for up to 18 percent of their inputs. Today, local firms supply a very limited range of CM inputs—cardboard boxes and shipping labels, cables and wires, and disposal services. In an interview with Flextronics, we discovered that the company even contracts with foreign trucking firms for transport services. According to our interviews, the major CMs operating in Guadalajara as of 2006—Solectron, Flextronics, SCI-Sanmina, and Jabil Circuit—import more than 95 percent of their inputs from overseas.

In a comprehensive study on subcontractors during the peak period of 1996–97, Dussel (1999) cites previous estimates that suggest that 87 percent of all machinery in the sector is foreign and that 61 percent of firms conduct no R&D (and most do not plan to). In all, Dussel estimates that the value added by Mexican firms to total production is only about 5 percent.

The failure of CMs to source locally stems from many sources, including the failure by local firms to upgrade technology and skills—and the failure of flagships and CMs to help build local firms' capacities to increase scale and quality. Willingness to help build local supply capacities was more evident in earlier years. IBM's promising 1993 "Jetway" plan called for suppliers to be co-located with production facilities and envisioned that 80–90 percent of imported inputs would supplied by regional or national firms by 2000 (Dussel 1999). To become a supplier, IBM required that firms obtain ISO 9000 (and sometimes ISO 14000) certification. According to our interviews, in the face of cutthroat global competition, however, IBM scrapped the program in 2002 and contracted nearly 100 percent of its manufacturing to SCI-Sanmina.

Nevertheless, in these few areas where CMs contract with local suppliers, there has been some training. Rivera Vargas (2003) found that some of the CMs provided quality control training to carton and packaging suppliers. One endogenous supplier of cables and wires that we interviewed had been trained to meet ISO 9000 standards.

Stuck on the Low Rungs of the Value Chain

Rather than spill-overs to local firms through backward linkages, the golden age wrought the transformation of Guadalajara's IT industry into a foreign enclave dependent on imported inputs. With little investment in re-building the long-term capacities of local firms, the Guadalajara IT industry has little hope of moving up the global IT value chain.

An extremely limited amount of R&D is being conducted in Guadalajara by foreign flagships, foreign CMs and local suppliers, or between any of these three entities and local universities—and the limited R&D that exists is skewed toward operations and manufacturing. In all of our interviews with the large multinationals and CMs, we were told that no R&D is conducted by the private sector in Guadalajara, though at times in the past it had been. In the cases of IBM and Hewlett-Packard, R&D is conducted at headquarters in the United States.

An earlier study found that during the 1990s ten patents were registered by foreign electronics firms operating in Guadalajara. All of these patents however, were registered in their country of origin (Rivera Vargas 2002). The CMs by their very nature work only to client specifications—flagships tend to keep R&D in house. We did observe some limited R&D occurring in one of the cable and wire suppliers that serves as a supplier to IBM.

In a study of industry-university relationships in Guadalajara's electronics industry, Rivera-Vargas (2002) found that eight of the region's 60 firms were involved in some type of collaboration with area universities. However, foreign firms were working with universities not to enhance product and process design capacities but to improve quality control and perfect assembly procedures for existing plants. Rather than promoting technological upgrading, she concludes, troublingly, that university-firm partnerships have helped to "lock in" the region as an assembly site:

In the case of Guadalajara, foreign investment swept away entrepreneurial capacity by pushing the endogenous electronics and computer industry out of the market. Second, foreign investment has limited the process of building scientific and technological capacities in the host country by demanding major emphasis in operational and manufacturing capabilities in detriment to the electronic design orientation (Rivera-Vargas 2002, p. 171).

Backward Linkage Survival Story: Electrónica Pantera

A handful of firms survived the tidal wave of MNCs in Guadalajara. The experience of one firm—Electrónica Pantera—serves as a crucible to view the difficulties of working as a supplier to the flagships.

Electrónica Pantera is a Mexican-owned firms that specializes in supplying cables and harnesses to IBM and Hewlett-Packard in Guadalajara, the United States, and Europe. Pantera's CEO, María Luisa Lozano, is a chemical engineer trained in the United States who worked in California for several years before returning to Guadalajara in 1974. There she went to work for Unisys, a spin-off from Burroughs that produced harnesses and magnetic heads for IBM, Burroughs, Motorola, and Kodak. Now defunct, Unisys had 2,500 employees at its peak. While at Unisys, Lozano met Bob Verbosh, her future husband, who was the sales manager in the Los Angeles branch of Burroughs.

In 1975, Lozano and Verbosh left the company and, with another five people from Burroughs, founded Pantera. Relying on their contacts and reputation as suppliers, as well as Mexico's local content requirements and high import taxes, they started the company with no initial capital other than some machinery and cash savings. Pantera supplied magnetic heads, cables and harnesses for Hewlett-Packard Texas and IBM Guadalajara and formed an association with Madison Wire Cable.

In 1996, Pantera entered into a joint venture with the US firm JPM, and the company's name was changed to JPM Pantera. Under the arrangement, Pantera held 40 percent of the firm's capital and controlled all plant-level operations. As the sector became more consolidated, JPM bought the whole firm in 1999. A new manager arrived, but according to Lozano it took him only months to "kill the company." By the beginning of 2001, JPM went bankrupt. In December of that year, Lozano and Verbosh bought back the company. The family now owns 100 percent of the plant.

JPM Pantera has recovered slowly from the bankruptcy. Whereas the firm once had 2,500 employees and three plants, it now has only one production site, with 500 employees. In 2002 it produced only cables and harnesses. "We came back to be a humble producer," says Lozano. Faced with stiff competition from China, Pantera survives on strict quality control and design flexibility, a solid reputation, and fast delivery to the US market.

By 2005, Pantera's main strategy was to keep quality, flexibility, and service at the highest possible levels. Lozano put it this way: "We need to find the profile that matches perfectly to the client's demand, no matter what." During our first interview with Lozano in October 2003, Pantera was celebrating its ninth month of zero defects and on-time delivery in its shipments to IBM Rochester.

A central problem in supplier development is the very low profit margin. Pantera's profit margin was 1 percent in 2003. By 2005, Pantera was feeling even more squeezed; regardless of past service, all the flagships and CMs had switched to global open-market bidding on the Internet for contracts. But the company is determined to hold on. "We won't go down without a fight," says Lozano.

Spill-Overs of Human Capital

Even if they do not have backward linkages to local firms, foreign firms can potentially generate positive spill-overs for local economic development through their employee hiring and training policies. Spill-overs of human capital occur when MNCs hire local workers and managers and provide opportunities to upgrade their skills on the job or through training programs. If and when such employees find jobs in local firms, bringing their newly acquired skills with them, they can increase the productivity and competitiveness of local firms. Similarly, human-capital spill-overs are captured when employees leave MNCs to form their own new "spin-off" firms.

Although there are a few interesting exceptions, on the whole there have been few examples of such human-capital spill-overs in high-tech Jalisco, either during the golden age or subsequently. Foreign IT firms in Guadalajara occupy a relatively low-skilled niche in the global electronics value chain. The thousands of workers employed by CMs to assemble IT products are generally low-skilled and receive only minimal initial training.

The educational distribution of work in Guadalajara electronics firms reflects the low skill level. The most recent studies estimate that 66.1 percent of all workers in the plants have the equivalent of a middle school education or less. Only 6.9 percent of employees have graduated from high school, and only 0.52 percent have post-graduate training. However, 100 percent of all employees in Guadalajara's plants are

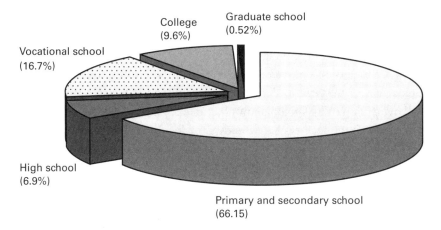

Figure 6.1
Distribution of educational levels in Jalisco electronics employment. Source:
Partida 2003.

Mexican (Partida and Moreno 2003). In one interview with a contract
manufacturer, we were told that during the IT boom the CMs hired buses
with loudspeakers to go out to villages and round up workers.

Another aspect of the employment process in Guadalajara that makes
spill-overs more difficult is the shift toward subcontracted employment.
Because wages in the region began to creep up relative to foreign competitors, the majority of workers in the sector (72 percent) are hired and paid
by the more than 25 employment firms in the region. Sixty-eight percent
of the subcontracted workers receive all their training at the employment
firm, not from the high-tech firm itself. The nature of these subcontracts
is very temporary, and accentuate turnover rates. Sixty-four percent of
all workers are women, and 80 percent of all workers are under the age
of 30. (In California's Silicon Valley, 67 percent are over 30.) Sixty-five
percent of all subcontracted workers have contracts for 1–3 months, 17
percent have contracts for 3–12 months, and only 18 percent have contracts for more than 12 months (Partida 2004).

Since the wipe-out of local electronics firms, there are very few firms
for the limited number of higher-skilled workers to spill over to. In our
interviews with workers and employees, we learned that they are more
apt to move from one foreign CM to another. Going to a local firm does
not cross their mind.

Spin-Offs: Software and Design

Although the general trend suggests a paucity of human-capital spill-overs, a small but burgeoning software and design industry has emerged in Jalisco. Much of this movement traces back to when IBM first established operations in the region. Remember that in 1985, Mexico granted IBM full ownership of its Guadalajara plant in exchange for a commitment to set up a training facility that would boost local technological capacities. Although IBM no longer participates in such partnerships, this one initiative has led to spin-offs that continue to spawn interesting developments 20 years later.

In partnership with IBM, the Mexican government established the Centro de Technologia de Semicontuctores. CTS opened on November 18, 1988 at the Centro de Investigación y Estudios Avanzados, National Polytechnic University in Guadalajara. The original idea was to do develop capacities to sell design services to Mexican-based foreign and domestic companies. However, it became apparent in a few years that Mexican-based companies could not use design services because they were all manufacturing companies and all of their design was done in the United States. Also, Mexican industry more broadly was not open to these kinds of services.

During the first 3 years, IBM provided the Centro de Technologia de Semicontuctores with full-time consultants and advisors who trained Mexican engineers to design integrated circuits. The CTS was never part of IBM. The contribution of IBM was just to train the engineers and provide technology for the center. After NAFTA, IBM ceased participation in the project. When all was said and done, IBM had probably provided about US$5 million, the bulk of which went to tools (computer programs, machines) and training. IBM's participation ended in 1995.

Despite a rather meager investment, the Centro de Technologia de Semicontuctores generated a handful of successful start-ups. We inter-viewed Jesus Palomino Echartea, currently general manager of Intel Tecnologia de Mexico's Guadalajara Design Center. Mr. Palomino is an engineer that was trained at CTS. Before working at Intel, he was the founder and CEO of Telecom Datacom Communications (TDCOMM), a local group of 32 people specializing in telecommunications that was 100 percent Mexican owned. TDCOMM had expertise in developing testing environments and had its own intellectual property in developing text scripts and text environments. They were consulting to US companies in New Jersey and in Connecticut. Their work attracted the attention of Intel Venture Capital, who acquired them in 2000. All the staff members have stayed on.

Several of the staff of TDCOMM were originally connected to the Centro de Technologia de Semicontuctores. Palomino reflects that CTS lasted as a joint venture from 1988 to 1994 (and still runs today without IBM involvement). According to Palomino, the main benefit of CTS was the exposure of Mexican engineers to technology—specifically, how to develop integrated circuits and computers. With IBM's involvement, CTS trained about 60 engineers. At peak, they had 32 people, but there were generally 25 engineers in training at a given time. Mr. Palomino himself spent nearly 6 years with CTS. Many engineers left CTS to seek engineering jobs in the United States. However, Palomino remarked that three firms spun off from CTS: TDCOMM, Mixval, and DDTEC. In 2004, only Mixval was still in operation.

In a recent further effort to develop endogenous productive capacity in the region's software sector, Palomino and others have formed Red de Ingenieros (REDI), a network of engineers focused on promoting the "spirit of association" in the electronics industry in Guadalajara. REDI was officially launched in 2003. The idea for the organization arose when it was clear that Guadalajara's electronics industry cluster was at risk due to the low cost of manufacturing in China. Local engineers started getting together to promote the idea of "created in Mexico" by keeping manufacturing in the region and/or helping develop other clusters to take the place of jobs/companies that may move to China. Eventually REDI will reach out beyond engineers (i.e. marketing professionals, attorneys, investors in the industry); however, since REDI's leaders are members of the engineering community themselves, the most obvious first step was to reach out to that network. They are also working with a group of Hispanic investors in California's Silicon Valley to attract Hispanic venture capitalists into their firms.

Building on the CTS training and the region's universities, a small but burgeoning number of software firms have emerged in Guadalajara.

According to CADELEC, Guadalajara houses 17 software firms, six of which are foreign and the rest Mexican owned. Recently the national Ministry of Economy launched a Program for the Development of the Software Industry (PROSOFT), whose aim is to reach a production level of $5 billion or for software production to reach the 4.3 percent of GDP OECD average (Mexico currently lags far behind that level at 1.3 percent). The elements of the plan, which is still under development, are to attract software FDI, strengthen the linkage between software firms and universities, develop a local market, and provide financing schemes for accompanying local firms.

Forward Linkages: Domestic IT Markets

The diffusion of computers and other IT products in industry and society has enormous potential to spur further innovations and products that promote economic growth and development. Through such forward linkages, the development of an IT sector could potentially have been a driver of productivity and growth not only in Jalisco but also in Mexico as a whole. Moreover, the growth of the domestic IT market could drive innovation in product niches in the Mexican IT industry. It is exactly this kind of "virtuous circle" of domestic market growth and domestic production that the earlier ISI strategy aimed to promote.

The liberalization strategy had a different aim: to increase local productivity by increasing IT exports, primarily to the United States. Computers and other IT products manufactured in Guadalajara did not enter the Mexican market directly. From a peak of nearly 60 percent of domestic demand in 1986, Mexico-based firms supplied only some 5 percent by 2001 (figure 6.2). With little linkage between domestic production and domestic markets, it is hard to detect any direct forward spill-overs.

Still, forward spill-overs could have been obtained indirectly if lower cost/higher quality computers and other IT products—manufactured in Mexico or elsewhere—were available to and widely bought by Mexican firms and consumers. Unfortunately, the story unfolded differently: the rate of domestic diffusion of IT products stabilized after liberalization.

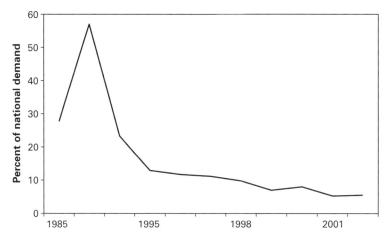

Figure 6.2
IT sales by Mexico-based firms. Sources: Peres 1990; National Accounts of Mexico, INEGI 2004.

In 2001, the sale of IT products as a percentage of GDP in Mexico was 3.2 percent, whereas in 1995 it had been 3.7 percent. While the number of personal computers per 1,000 people in Mexico doubled between 1995 and 2001, it tripled in Chile and grew by more than 350 percent in Brazil and Chile. Indeed, with 50.6 computers per 1,000 people in 2001, Mexico fell well below the Latin American average of 59.3. In 1995, only Chile had a higher rate of diffusion than Mexico. The trends are similar for the diffusion of telephone mainlines and mobiles. In both areas, Mexico fell well below the Latin American average in 2001 (table 6.2).

Some firms and the Mexican government recognize that Mexico's domestic market is falling behind and have set up programs to bridge the digital divide. In 2001, IBM worked with the Technical Institute of Studies of Monterrey to provide Learning Village, a Web-based tool for teachers, in Spanish and to provide training for teachers to learn the tool. In 2003, IBM and the state of Hidalgo initiated a program to improve secondary school instruction using Learning Village to share information among teachers. IBM is also helping promote the use of community digital centers (IBM 2004).

In 2001, President Vicente Fox announced a plan to wire 98 percent of Mexican communities by 2006. The goal of the Sistema Nacional

Table 6.2
IT diffusion in Western Hemisphere. Source: World Bank 2004.

		Telephone main lines per 1,000 people	Mobile phones	PCs	IT/GDP
United States	2001	667	452	625	7.9
Latin America	2001	163	160	59.3	n.a.
Mexico	1995	94	7	25.6	3.7
	2001	125	142	50.6	3.2
Brazil	1995	85	8	17.3	2.7
	2001	218	167	62.9	8.3
Chile	1995	127	14	33.3	4.2
	2001	233	342	106.5	8.1
Argentina	1995	162	10	24.4	3.6
	2001	224	193	91.1	4

e-Mexico program, known as e-Mexico, is to build a more technologically sophisticated country and to reach out to the corners of the country via the Internet and a host of government services provided online. The program has a Web portal that it meant to serve various public needs (learning, government, health) and provides specialized content to a range of interest groups (women, children, visitors, students). However, since its inception in 2001, e-Mexico has been largely focused on the education necessary to make the e-Mexico goal a reality, training teachers to use technology and teaching schoolchildren how to use the Internet.

In 2002, Fox's government announced it had completed the R&D phase of the project and was ready to move onto implementation with a 667 million-peso budget and a private investment goal of US$4 billion. In 2002, commitments from the private sector included US$17 million from Intel to provide computer training for elementary school teachers, and a five-year commitment by Microsoft to provide software (via educational licenses) and support for 250 public Internet kiosks called Telecentres. The donation, which Microsoft valued at US$6 million, includes the training of 4,000 technicians to help run the kiosks. Mexico's eventual goal is to set up 10,000 such kiosks throughout rural communities in the country (Jacobs 2002).

Open source advocates and members of the Mexican legislature have argued that Microsoft's intention was to stifle innovation by committing the government to Microsoft's proprietary software (Gori 2002). Indeed, Wade (2002) warns that the promotion of information and communication technologies as a key development initiative runs the risk of "locking developing countries into a new form of dependency on the West." Moreover, unless firmly rooted in local needs and economic realities, such initiatives stand little chance of success. Of the 23 Telecentres that were installed, only five remained a year later.

Mexican Policy: Tilting toward Foreign Firms

The golden age of Guadalajara's "silicon valley" not only delivered few spill-overs to local firms but destroyed many. While there are multiple reasons, Mexican government policy—both in commission and omission—played a decisive role. Given the change in the global organization of the IT industry, policy had to be especially strategic and proactive in order to work with the MNCs to nurture domestic capacities for production and innovation. Rather than promote domestic firms or even create a "level playing field," the overall effect of Mexican industry, credit, foreign exchange and especially trade policies was to create a bias toward foreign firms.

Industry and Credit Policies

There are two broad policy approaches to building domestic firm capacities. The first involves policies that enhance the general climate for learning, innovation and investment throughout the economy, including support for transport and communications infrastructure, secondary and tertiary education, science and engineering education, a comprehensive science and technology policy, worker training programs, etc. Credit programs providing domestic SMEs in all sectors with access to credit at market rates also fall in this category. The second approach involves policies targeted at building domestic firm capacities in specific industries—for example, support for R&D; targeted access to credit (either on market terms or subsidized); targeted training programs; and brokered partnerships between MNCs and domestic firms and among MNCs, domestic firms, and universities. Also in this category are performance requirements. Such requirements, such as that MNCs should enter into joint

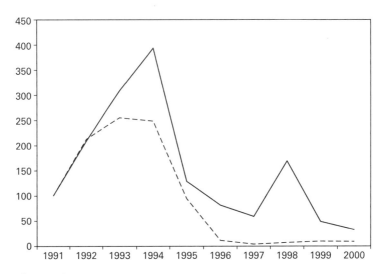

Figure 6.3
NAFIN credit to SMEs in Jalisco. —: millions of pesos. --: number of recipients.
1991 = 100. Source: SEPROE 2001.

ventures with domestic firms, should export a portion of their product, or should source a portion of their inputs domestically, were used intensively by East Asian countries to build up their IT industry in the 1970s and the 1980s. The governments of Mexico and Jalisco undertook some of the first set, though arguably they could have done much more, especially in the areas of science and technology policy and support for R&D. However, they undertook none in the second set. The three bodies working to promote the IT industry in the region at a modest level—CANIETI, the State of Jalisco, and CADELEC—fell short of effectively helping local producers. For CANIETI, the origin of suppliers is irrelevant: the goal is to keep foreign firms in the country and attract new ones. Moreover, CANIETI was poorly funded and poorly coordinated. The State of Jalisco was (and is) also focused on attracting foreign firms, including through the provision of tax breaks. Charged with assisting the incorporation of local suppliers into flagship and CM supply chains, CADELEC makes little differentiation between of locally based foreign CMs and small and medium-size local domestic firms.

In their detailed study of the collapse of the IT industry, Dedrick et al. (2001) conclude that the lack of a strategic industrial policy was the primary reason for the failure of Silicon South. They found that there was very little support by CANIETI, by CADELEC, or by the State of Jalisco for clear skill upgrading policies in higher education and on-the-job training; for support for local firms through low-cost loans, export assistance, and business incubation for more support for university-industry cooperative research and coordination on business incubation; fortele-communications infrastructure improvements to increase availability and quality of IT services; or for more targeted incentives to match suppliers with TNCs in the region.

With razor-thin profit margins, access to credit is especially important for SME suppliers in the IT industry. Mexico's domestic financial markets, however, are largely dysfunctional, leaving SMEs with little or no access to credit. SMEs do have access to NAFIN, Mexico's national development bank. Once a powerhouse of the ISI strategy, NAFIN's budget was severely restricted in the liberalization era. Both the total amount of credit and the number of SME loan recipients in Jalisco declined dramatically between 1994 and 2000 (figure 6.3). Moreover, Mexico suffered peso and financial crises in the 1990s, cutting off or restricting access to other sources of credit. The only external financing option available was in foreign capital markets—an option not open to domestic firms.

Trade and Foreign Exchange Policies

Mexico's macroeconomic policies, especially trade and foreign exchange, heavily tilted market dynamics toward foreign companies. The PITEX program, for example, explicitly provides incentives to import intermediate inputs rather than purchase them locally. Under PITEX, imported inputs are duty-free as long as more than 65 percent of finished product is exported (Dussel 2003). NAFTA offers even lower tariffs. While the program is open to both foreign and domestic firms, in practice, it is the foreign firms who are in the position to export. As the economist Amartya Sen has remarked, "equal rules for unequal partners constitute unequal rules."

PITEX's double incentive—to import inputs and export products—is doubly detrimental from a spill-overs perspective. Not only does it encourage MNCs to overlook local suppliers, thus cutting off the potential for backward linkages; in its tilt toward exports, it reduces the scope for forward linkages through sales in domestic markets.

The ability of the Mexican government to develop and implement a strategic industrial policy for the IT sector was further restricted by the macroeconomic policies needed to support NAFTA and the liberalization policy as a whole. Two policies were maintaining high interest rates and keeping the peso strong relative to the dollar.

High interest rates and an overvalued peso aimed to control inflation, thus attracting foreign capital and inhibiting capital flight. Mexico suffered financial crises twice in the 1990s and the specter of rampant inflation is ever-present, especially in the context of Mexico's external debt and continuing current-account deficits. The peso crisis that rocked the country in 1994 hit consumers and small and medium-size firms the hardest.

Exchange rate and interest policies squeezed SMEs in Guadalajara's IT sector in two ways. The overvalued peso provided an additional incentive to flagships and CMs to import intermediate inputs rather than purchase them locally. Besides contributing to peso overvaluation, high interest rates choked off domestic investment. Foreign firms, including in the IT sector, had access to internal company sources of finance and global capital markets. Domestic firms did not.

In order to have taken advantage of the flagships' shift to outsourcing in the mid 1990s, Jalisco's local firms would have had to invest in expanding their scale of production. Developing scale capabilities entails large sunk costs. Given that Mexico had just gone through a severe economic crisis, SMEs did not have the cash internally and many went out of business. Even if the flagships considered the possibility of contracting with local firms when they began outsourcing manufacturing functions, it was not possible to find many who could ramp up to scale. The ability to manufacture at high scale was found to be the key variable for Taiwanese firms that went from third tier assemblers, to CMs and eventually to flagship firms in their own right (Amsden and Chu 2003).

After the peso crisis in 1994, the Mexican firms we interviewed told us they spent so much effort trying to survive that there was little time or money to think about the far future. The macroeconomic environment, in other words, provided a disincentive to "learn" how to produce at high scale (that is, to conduct R&D, develop supply linkages and marketing opportunities and so forth). The problem afflicted the Mexican economy as a whole. In an atmosphere of imperfect competition and macroeconomic uncertainty, local firms tend to act conservatively (Cimoli 2000).

Even if they were inclined to be aggressively entrepreneurial, local firms were hamstrung by the squeeze on domestic credit. Between 1994 and 2003, total domestic credit as a percentage of GDP fell by 66 percent. Even more dramatically, credit for firms decreased by nearly 84 percent. In a survey of Mexican firms, the Bank of Mexico learned that an astonishing 82 percent were not able to access credit between 1998 and 2003. The majority were squeezed out by high interest rates, while the rest were simply rejected (UNIDO 2005). With access to internal sources of finance, as well as global capital markets, the MNCs were exempted from the credit squeeze.

Could Guadalajara have evolved into a dynamic knowledge-intensive IT cluster with more institutional and policy support from state and federal governments? Given the cutthroat competition in the global IT industry and the vulnerability to lower wage producers like China, success—even with government support—is not guaranteed. What seems clear, however, is that only within a more strategic, proactive policy context did Mexican firms have any chance to develop supply linkages to and capture spill-overs from MNCs. Without such local capacities, MNCs lacked a compelling reason—save for proximity to the United States—to develop productive functions in Guadalajara. When global market conditions changed, they evacuated.

7

Importing Environmentalism?

For Mexico, as for other developing countries, foreign investment in IT promised the growth of an industry with safe working conditions, a cleaner natural environment, and cutting-edge environmental technology. Compared to Mexico's traditional smokestack manufacturing industries such as chemicals, petroleum, mining and leather tanning, the IT industry seemed a "clean and green" alternative. Besides inherently cleaner production processes, foreign IT companies would bring with them "best practice" environmental management systems, the benefits of which would spill over to local manufacturing and supplier firms.

But the IT industry is far from "clean and green." (See chapter 3.) As is true of most developing countries, Mexico's capacity for the regulation of risks to workers, public health, and the environment are thin. In the 1990s, the IT industry in Guadalajara blossomed largely within a regulatory void. When it came to managing impacts on the environment and occupational health, multinational corporations virtually governed themselves. Moreover, with the crowding out of the domestic IT industry, few Mexican firms were in place to capture any potential spill-overs. Nevertheless, an interesting coalition of international non-governmental organizations and Guadalajara-based labor and environmental groups formed to put pressure on foreign IT firms to clean up.

This chapter considers whether high-tech FDI generated sustainable development spill-overs in Guadalajara.

Environmental Regulation

Like other developing countries, Mexico has a limited capacity to regulate industry. Not all industries receive equal attention, either in the passage of environmental laws or their implementation. Electronics generally and high tech in particular are not priority areas. At the state level, emerging environmental laws generally ignore manufacturing altogether and focus on forestry and water issues.

With the exception of broad rules applicable to all industries, such as the requirement to dispose of hazardous wastes and sulfur dioxide emissions off-site, high-tech industries operate in a "lax policy environment" (ECLAC 2003). Like all manufacturing industries, the IT sector is subject to a number of *normas oficial Mexicanas* (Official Mexican Norms) for environmental protection. Among them are maximum limits for contaminants in residual waste for water and national goods, maximum levels of atmospheric emissions of sulfur dioxide and trioxide from plants that emit sulfuric acid, and maximum levels of atmospheric emission of solid particulates from fixed plants.

Another regulation, referred to as the dangerous waste norm, establishes environmental characteristics of dangerous waste, including corrosive, reactive, explosive, toxic or inflammable substances (Nom-052-Ecol-1993). The norm states that dangerous waste should be "handled and disposed of adequately so as not to affect human beings or their environment."

One of the general "normas" most relevant to the IT industry is the requirement that firms taking part in the maquiladora and PITEX programs must return toxic waste of production processes to their countries of origin. Since US companies dominate in the high-tech sector, this means shipping toxic waste across the border.

The industry also has to comply with the General Law of Ecological Balance and Protection of the Environment (LGEEPA) formed in 1988 and modified in 1996. One clause specifically addresses maquiladoras and dangerous waste, reinforcing the return to country of origin rule. There are also commitments under the environmental side agreement of NAFTA to eliminate two toxic substances used by the electronics industry: mercury and lead.

Beyond these general laws governing hazardous waste, there is no industry-specific environmental or health and safety regulation of IT manufacture and assembly in Mexico. Moreover, Mexico's environmental protection agency, PROFEPA, is stretched thin when it comes to enforcing even these laws. In recent years, resources devoted to industry inspection have been falling. In 2002, PROFEPA made only 445 inspections of maquiladora companies on Mexico's northern border, down from an average of 1,000 per year between 1994 and 2000. Some 60 percent—274 companies—received infractions, while 170 had no infractions and one was closed temporarily (Stromberg 2002).

The electronics industry in general and the IT industry in particular operate with little oversight and a great deal of opaqueness as to their compliance with environmental laws, including in the all-important matter of the management of electronic waste generated in production.

There are two kinds of electronic waste from IT production: liquid wastes and scrap. The scrap consists of solid waste and hazardous waste. According to our interviews with PROFEPA, the agency has certified one firm, Environmental Electronics Recycling, to take and separate the scrap from IT manufacturing sites. The solid waste is sent to Japan, while the hazardous waste is sent to a treatment plant in Nuevo Leon (the only hazardous-waste-treatment plant in Mexico) or to the United States. At least, that is the requirement. Although permits are needed for every transaction along the chain, PROFEPA has a very limited monitoring capacity. In the little monitoring that they have undertaken, they found a significant amount of fraud, including forged permits. When the fraud was uncovered, the waste-hauling firms simply disappeared.

How much of the hazardous electronic waste actually makes it to the United States? Separate data for the electronics industry is not available. One study found that, of the 60,000 tons of waste that required special handling generated in 1996 by Mexico's maquiladora industry, 60 percent was returned to country of origin and 12 percent was kept in known locations in Mexico. However, the whereabouts of 26 percent of the waste were unknown (ECLAC 2003).

Moreover, many firms do not report the extent to which they are using hazardous waste in the first place. A recent study reports that only 30 percent of all industries required to report their hazardous waste use actually do report. Of the IT firms in Guadalajara, only one firm filed notices of hazardous waste return to the Mexican authorities—SCI-

Semina, which was in the Industria Limpia program (TCPS 2004). Even when IT firms do report their use of hazardous substances, such reporting is incomplete. Citing the protection of trade secrets, IT firms routinely do not allow PROFEPA into IT production "clean rooms."

The problem of the "disappearing waste" is exacerbated by the lack of good data not only in Mexico but also in the United States and Canada. A 2004 study of trans-boundary hazardous waste in NAFTA found that that "return" waste from Mexican maquiladoras overall seemed to be increasing, due to better compliance and an improved reporting system in Mexico. However, there are no tracking systems for the category of electronic waste in Mexico, the United States, or Canada (TCPS 2004).

Another gap in the Mexican regulation of risks posed by industry is in the area of occupational health and safety. Foreign circuit board assembly operations based in Guadalajara use a considerable amount of lead in attaching components to copper plates. Moreover, the copper plating process emits formaldehyde and brominated flame retardants, posing health hazards to workers. The on-site handling and storage of lead and the transport and "disposal" of hazardous wastes are important environmental and health issues not only for workers but also for local communities (Kuehr 2003). As of early 2004, Mexico had no OHS standards for worker exposure to lead, formaldehyde, or brominated flame retardants.

While there is little oversight of IT production, Mexico's new waste management law of 2003 requires electronics firms to "take back" products. When fully implemented, the law will require IT companies to take back used products, returned products and those that were produced and not sold. IT companies will have to find local firms who will strip electronic products for precious metals and hazardous materials.

The Industria Limpia Program

Given the difficulty in wielding the "stick" of inspections and fines, PROFEPA devised an innovative, voluntary "carrot"-based certification approach to improve environmental compliance in manufacturing industries—the Industria Limpia (clean industry) program. The certification is given on a facility, not company-wide, basis, though many firms operate only one plant in Mexico.

When a firm signals interest, PROFEPA sends an auditor to the plant to do a formal audit of the firm's environmental activities. The firm chooses an auditor from a list of private sector certified auditors. After the audit is completed, PROFEPA negotiates an action plan with the firm based on improvements toward quantifiable targets. One feature of the program especially relevant to the electronics industry is that all firms must register on-site hazardous wastes with PROFEPA.

When the firm signs an agreement to adhere to the action plan, it is officially admitted into the program. The firm must demonstrate concrete improvements on an annual basis and meet specified targets. If they do not, they face the inspections and fines required by law for non-compliance.

Once the firm has achieved compliance, it gets "certified" and can use the Industria Limpia label to promote their products in domestic and foreign markets. Certification expires after 2 years. To stay in the program, firms must quantifiably demonstrate every 3 months that they remain in compliance. If firms want to go beyond compliance they can be given a certification of "excellence."

With its firm-specific agreements, this program is designed to attract the "leader" firms in corporate social responsibility in hopes they will lead by example. In Jalisco, the main focus of the Industria Limpia program has been not electronics but chemicals, leather and molding industries. The only IT firms to date that signed up are SCI-Semina, IBM, Hewlett-Packard, and Motorola, and only SCI-Semina completed the process all the way to certification. The problem may be that the benefits of the Industria Limpia label are unclear. For companies selling into the European market, what matters is compliance with EU, not Mexican, standards. Likewise in Japan and in the United States.

According to PROFEPA, lack of interest stems from the fact that the requirement to quantifiably demonstrate full compliance is difficult and costly. Certification to Industria Limpia requires adherence to an environmental management plan developed outside the company and monitored by external auditors. By contrast, certification to ISO 14001 allows companies to set—and monitor—their own targets. SCI-Semina achieved ISO certification in only 3 months but took 2 years to get the Industria Limpia label.

Even so, Industria Limpia certification does not guarantee safety to workers or the community. In May 2004, there was a toxic explosion at SCI-Semina just a few months after it was certified. A plate fell off a solder machine and began to burn noxious gases. The 400 people working on the shift were evacuated for an hour. According to local newspapers, 39 employees were taken to the hospital for respiratory or nervous system problems (Zamora 2004). Some reports also indicate that, despite its ISO and Industria Limpia certifications, the firm was not prepared for the incident and could have prevented the hospitalizations by keeping the plant shut longer. Out of fear, fifteen workers resigned from the plant. One worker told us: "They made an error in making us return to our posts too soon—hardly an hour after the cards fell—and this is what complicated things. They had increased the power in the air vents, but even so, about twenty people began to feel sick, get dizzy, [and] throw up. So, we were evacuated again."

One of the serious shortcomings of both ISO 14001 and Industria Limpia is the lack of specific measures for worker health and safety. Although ISO has developed the 18000 series for occupational health and safety, few Mexican-based companies have implemented it. According to one recent study based on interviews with a large sample of workers in Guadalajara IT plants, workers are not well-informed about the environmental risks of their work. The study reports that 45 percent of IT workers do not know whether they are handling substances that are deemed to be toxic. Seventy-nine percent of those that do know they are handling toxic substances do not know how or where the waste is disposed. The study also found that 19 percent of all workers report that they have been sick on the job and 7 percent have experienced an environmental accident (Partida 2004).

Pollution Halo?

There is little doubt that environmental regulation of the IT industry is less stringent in Mexico and other developing countries than in the United States, Europe or Japan. With rising public concerns about environmental and health problems, did lower environmental standards attract MNCs to invest in Guadalajara? In other words, was Guadalajara slated to be a "pollution haven" for the global IT industry, especially US companies?

There is little evidence that environmental standards played a role in the decision of US manufacturing firms in general to locate in Mexico after the passage of NAFTA. (See chapter 2.) Based on our interviews with flagship companies, as well as published reports, we conclude there is no evidence to support the "pollution haven" hypothesis in the case of the IT industry. Investment in Guadalajara was driven by the large expansion of the US market in the 1990s, in turn propelled by falling product prices and the high-tech stock market bubble. IT companies came to Guadalajara because of the low labor costs and the proximity to the booming US market (chapter 5). Environmental standards simply were not on the radar screen.

Rather than why they came, the more pertinent question is how MNCs behaved in Guadalajara once they got there. Given low standards and lax enforcement, both foreign and domestic companies have a wide scope to self-regulate—to design and implement their own environmental and occupational health management strategies and standards. Part of what makes FDI attractive to developing countries is the hope that MNCs will create a pollution halo through the transfer of clean technology and best practice environmental management (Ruud 2002). The expectation is that MNC corporate parents have global policies that mandate such transfers to their developing-country affiliates.

But not all MNCs adopt global, company-wide environmental standards. Some direct their foreign affiliates to follow host-country standards or, if standards are absent, local practice. The result is that, from the Mexican vantage point, an MNC may have double standards: one set for the home country, the other set for Mexico (and other host-country locations). Moreover, among those MNCs that do set global standards, there are differences in the rigor of monitoring. In this context, the environmental performance of any particular or group of MNCs cannot be predicted: it could or could not transfer clean technology and "best practice" and its environmental performance could be better, the same, or worse than domestic firms.

There is little data and few studies on the environmental performance of MNCs in Mexico's IT sector. A World Bank study that analyzed the

determinants of environmental compliance in Mexican manufacturing
as a whole found that foreign firms were no more likely to be in compliance than domestic firms (Dasgupta 2000). In 2002, the UN Economic
Commission for Latin America and the Caribbean conducted a survey of
298 electronics firms on Mexico's northern border with the United States.
The study found some signs that environmental compliance was improving. About two-thirds of the companies in their large sample—more than
three-fourths of which were foreign—had an environmental unit. More
than half said they had raised the level of environmental protection in the
preceding 3 years.

However, the report identified what it calls a foreign firm "double
standard" by showing that the country source of FDI had no influence
on the environmental policy or performance of the plant. Only half of
the companies had an "active" environmental policy, defined as environmental management measures complemented by technologies to mitigate
contaminant emissions from the plants. Moreover, only half had even
limited supervision of Mexican applicable environmental laws, which
are much lower than norms in other countries with electronic assembly
industries, such as the Philippines.

The report cautioned that this low level of performance would
create barriers for Mexican products in markets with high environmental requirements, such as the European Union (see below). Comparing
Mexico with the Philippines, the report concluded that, overall, the
Mexican electronics industry is being left behind in the speed and scope
of environmental policy, especially in terms of production processes and
product life cycle (ECLAC 2003).

Environmental Performance of Flagships
There are no published studies of the environmental performance of IT
firms in Guadalajara. To gain some insight, we interviewed all of the
major US flagship and all but one of the major CM companies, as well
as a handful of local supplier firms, government officials, and non-governmental organizations. We also examined company websites and other
reports.

The five flagships we interviewed in Guadalajara all have in place
an environmental management system (EMS) and a professional staff

Table 7.1

Environmental management by flagship OEMs in Guadalajara. PFC stands for perfluorocarbons, which are used in semiconductor manufacturing and are characterized as greenhouse gases because they are long-lived in the atmosphere. (Lucent had manufacturing operations in Guadalajara as of early 2001.) Source: interviews by the authors.

Company programs/targets	Hewlett-Packard	IBM	Lucent	Motorola	NEC
Energy efficiency/Greenhouse	PFC* emission reduction targets; energy efficiency targets	PFC and CO_2 emission reduction targets; energy efficiency target			PFC and CO_2 reduction targets
Lead and other heavy metals	Compliance with applicable regulations				Lead-free production (target)
Product stewardship	Planet Partners Recycling	Product Stewardship Program			
Water management		Water savings goal			
Supply chain	Social and environmental responsibility (SER)		Eco-environmental inquiries	Business conduct expectations Human rights policy	Green procurement guidelines
Worker health and safety	EHS management system				
Environmental management system	EHS MS; ISO 14,001	Worldwide EMS; ISO 14,001 surveillance audits	Internal	ISO 14,001	Internal EMS based on business line
Product design	Design for environment		Design for environment		
Transparency/disclosure	Global citizenship report	Corporate responsibility report			Annual environmental report
Internal worldwide company standards	Yes	Yes	No	No	No

charged with monitoring it (table 7.1). All are ISO 14001 certified, and some have gone beyond ISO to create their own EMS. Moreover, all the companies said that, during the period when they were conducting manufacturing or assembly operations in Guadalajara, they sent waste back to the United States through certified waste handlers.

We were not able to obtain facility-level data about the environmental performance of the flagships who currently or in the past undertook manufacturing and assembly operations. However, given different corporate cultures and policies, it is likely that performance varied. Hewlett-Packard, for example, is a recognized leader in global corporate social responsibility (CSR). In a comprehensive assessment, the European firm ISIS rated Hewlett-Packard as a "race leader" in terms of CSR in the IT sector. KLD, a US-based CSR research firm, says Hewlett-Packard (ISIS 2004) "has demonstrated a superior commitment to management systems, voluntary programs, or other environmentally proactive activities."

One problem is that such assessments, especially those that provide "triple bottom line" information for socially responsible investors, evaluate MNCs solely on the basis of available information in home countries. Few OECD countries require companies to report on their overseas affiliates. With its headquarters in the United States, Hewlett-Packard—like other US companies—has no obligation to disclose information about its foreign operations that is required of its domestic operations, such as toxic releases and transfers, air and water emissions, and occupational health and safety violations (Leighton et al. 2002).

There is some reason to think, however, that Hewlett-Packard may generally promote best practice in its affiliates in Mexico and elsewhere. Hewlett-Packard has adopted global standards as company policy and is ISO 14001 certified for manufacturing operations worldwide. The company operates with global- and regional-level management systems for EHS and in 2003 introduced Web-based data collection to strengthen their analytical and management functions. Site audits to ensure compliance to corporate EHS policies are conducted by internal specialists and, on an annual basis, both by the sites themselves and by a third party (Hewlett-Packard 2004).

One flagship in Guadalajara, Lucent Technologies, voluntarily went well beyond implementing an EMS or complying with Mexican law.

Lucent's state-of-the-art Guadalajara plant, built in 1991 and employing 12,000 workers at its peak, was the first facility in Mexico designed to have zero effluents. They also installed an innovative wave soldering process, which replaced manual soldering and facilitated capture and recycling of solder dross. The spent dross was shipped offsite for recycling to the United States. Water emissions were filtered and used for irrigation, rather than discharged into the local water system, an innovation sparked by the company's concern about NGO criticism.

Unfortunately, Lucent's state-of-the-art facility went bust—a victim of the technology stock crash in the early 2000s. Lucent's stock was one of the most overvalued of all the dotcoms, plummeting from its peak of $73 to $1. Once slated to be the "jewel" of Lucent's production in both north and south America, the Guadalajara plant was shut down and half of it was sold to an Asian firm who has since moved to China. As of mid 2005, the other half was still up for sale.

IBM has a mixed reputation. The company sets global standards for all of its affiliates (table 7.1). Since 1997, all of IBM's manufacturing and hardware development sites worldwide have been ISO 14001 certified (IBM 1997). In 2000, the company launched a computer take-back program for small businesses and individuals. But in the United States, IBM has been plagued by lawsuits from former workers charging that the company was responsible for their rare forms of cancer and other sickness. (See chapter 3.) The failure of the suits is generally attributed to the lack of health data, rather than the vindication of the company.

The ISIS report referred to above describes IBM as part of a "chasing pack" that has put in place some interesting environmental policies but has a long way to go to be an industry leader. KLD lists IBM's hazardous waste as a concern. This concern is raised when a company's "current liabilities for hazardous waste sites exceed $50 million, or the company has recently paid substantial fines or civil penalties for waste management violations."

Rather than routinely embracing and transferring "best practice," the picture that emerges is that flagships are evolving—at different rates—toward it. Moreover, "best practice" is not yet good enough even in developed countries to move flagships away from manufacturing processes that endanger human health and the environment. On May 23,

2005, Greenpeace activists demonstrated in front of Hewlett-Packard's plant in Guadalajara, calling for Hewlett-Packard and all other high-tech companies to eliminate toxic and hazardous substances in all its products. Part of Greenpeace's global campaign to reduce toxics in the high-tech industry, protests were held on the same day in Beijing and in Geneva (Greenpeace 2005).

Contract Manufacturers and the Environment

Little hard information is available about the environmental performance of CMs in Guadalajara. All generate significant amounts of waste, solder dross, lead, and tin. All have an EMS and an EHS unit, one has applied for Responsible Care certification and another is being certified to ISO 18000. Like the flagships, all CM firms told us they sent hazardous waste back to the United States through certified disposal services.

Our interviews and site visits revealed, however, that CMs have not fully grappled with environmental challenges, especially the exposure of workers to hazardous materials and toxic chemicals. In one plant we visited, the smell of solvents on the assembly floor was so overpowering that we had to leave after a few minutes. All CMs conduct blood tests for lead and urine to test for solvents at least twice a year. One firm uses the American Council of Government Industrial Hygienists standards that establish threshold limit values based on permissible exposure standards based on exposures of 8 hours per day, 40 hours per week, and 20 years of working.

Given that CMs favor temporary, usually 6-month contracts, benchmarking to such standards means that the companies simply avoid rather than grapple with the issue of worker health risk due to chemical exposure. For example, according to interviews with workers carried out by the UK-based Catholic Agency for Overseas Development, women workers are subjected to regular and intrusive questioning to determine if they are pregnant. If they are pregnant, they are subject to immediate job termination. The short term contracts disqualify them from maternity benefits (CAFOD 2004).

Since CMs work on contract to flagships, they take their cues on environmental management from their clients. Like their clients, environmental commitment apparently varies among CMs. Solectron, which denied our requests for an interview, considers itself an environmental leader. The Guadalajara site is certified to ISO 14001, as well as ISO 9001 (quality) and OHSAS 18001 (occupational health and safety). In 1999, only 2 years after it was established in Guadalajara, the company won the State of Jalisco's Environmental Award. "We won," said the President of Selectron de Mexico in 2000, "because we're totally committed and dedicated to follow the Mexican standards" (Montoy 2001). Solectron's most recent Corporate Responsibility Report, however, offers only one page on the environment, with very little substantive information about the extent or efficacy of any corporate initiatives (Solectron 2004). Given the comment above, it seems safe to assume that the company does not set or monitor global standards. Solectron's website describes a number of environmental sustainability programs—water recycling, materials recycling, lead-free electronics, environmentally friendly technologies— although it is not clear which sites participate in these programs and to what extent.

Jabil Circuit provides very little discussion of their global policies or practices. A 1999 press release announced Jabil's ISO 14000 certification for a Scotland plant, stating that all of its plants would be certified by 2001 (Jabil 1999). In our interview, EHS personnel said that the only federal laws that cover electronics in Mexico are the hazardous waste laws requiring registration of all hazardous waste used on site and the maquila program requirements to return hazardous wastes to their country of origin. They also said that the company does not use contract workers and monitors their regular workers three times a year for exposure to lead and solvents. According to our interviews, Jabil is considering joining the Industria Limpia program to improve their corporate image and because ISO 14000 recommends compliance with local laws.

Flextronics has had a lead-free soldering initiative since 1995 and has introduced various products using lead-free solder since 1997. The company's Director of Advanced Process Technology received the Soldertec Lead-free Solder Award in 2002 (Flextronics 2003). Flextronics sees itself as an environmental leader and is working to meet the new EU chemical and electronic waste regulations. (See chapter 3.) Following a directive by the corporate parent that Flextronics "look the same all over the world," the Guadalajara facility is certified to both ISO 14001 and

ISO 18000. Personnel from corporate headquarters visit the Guadalajara plant once a quarter to examine EHS (and other issues) and make sure it is congruent with corporate policy.

Environmental Spill-Overs to Domestic Firms?

Beyond the transfer of clean technology and good management to their own affiliates, MNCs can generate environmental spill-overs to domestic firms through demonstration effects and requirements placed on local suppliers. Given the absence of local manufacturing competitors, the primary channel for environmental spill-overs in Guadalajara would be via local suppliers.

Hewlett-Packard currently has a strong supply-chain policy. Established in 2002, its Supply Chain Social and Environmental Responsibility (SER) Policy commits the company to "[working] with suppliers to ensure they operate in a socially and environmentally responsible manner." In 2003, the company worked to ensure SER compliance with 50 suppliers, who accounted for more than 70 percent of Hewlett-Packard's procurement spending. In 2004, in conjunction with Dell and IBM, Hewlett-Packard took the lead in bringing together a coalition of eight flagship and five CM firms that created and launched an Electronics Industry Code of Conduct. The code outlines broad management standards for the electronics supply chain in the areas of labor rights, health and safety, and environment (Hewlett-Packard 2004).

The implementation of Hewlett-Packard's policy, however, may be spotty and may have missed Guadalajara, especially in the past. An EHS manager for Hewlett-Packard in Guadalajara told us that "Hewlett-Packard's theory is to keep costs down . . . and for that reason we do not make demands on suppliers" (Hewlett-Packard 2003). One SME supplier confirmed that none of their flagship clients, including Hewlett-Packard, had ever imposed any environmental requirements. However, Hewlett-Packard Mexico was part of a small business group formed in 1993 to promote best practice, including ISO 14001 certification, among manufacturing firms in Guadalajara.

IBM claims that it evaluates all of its hazardous waste disposal vendors and is committed to contracting with suppliers capable of managing "environmentally sensitive operations." It also encourages suppliers to become ISO 14001 compliant and will work directly with suppliers to develop more environmentally friendly processes when appropriate.

According to ISIS, however, the company takes a "hands-off" approach to suppliers: "IBM does not have an environmental supply chain policy although it has started to evaluate production-related supplier performance." When IBM identifies a weakness with a supplier, the supplier is expected to create a plan of action on its own. On-site suppliers have to comply with IBM's environmental policy (ISIS 2004).

During its heyday in Guadalajara, Lucent "mentored" its subcontractors to improve environmental performance. The US parent worried that imposing mandatory conditions on suppliers would generate liabilities. Along with Hewlett-Packard, IBM, Motorola, and others, Lucent created an industry association to promote common flagship strategies to reduce costs by developing local suppliers. Lucent also participated in and contributed funds toward a World Bank project in 1997 and 1998 which aimed to train SMEs in environmental management. The World Bank matched every dollar provided by the larger "mentoring" firms. However, the project's success was mixed. In some cases, the mentoring foreign firms themselves did not have an EMS, reducing their capacity to positively influence and work with their suppliers (World Bank 1998).

The policies and good will of at least some of the flagships suggested that environmental spill-overs, at least theoretically, would be on offer in Guadalajara. In the event, two factors account for the lack of such spill-overs. First, none of the CMs have procurement management policies requiring good environmental performance from their numerous global suppliers. Second and more important, as we showed in chapter 5, there were very few backward linkages to local firms in Guadalajara. The flagships contracted with the CMs—and the CMs sourced almost nothing locally.

Environmental Upgrading?

A third channel for sustainable development spill-overs is through technology leapfrogging—the upgrading of local environmental technology to the global frontier.

Environmental standards are set to rise dramatically in Europe with new EU legislation on hazardous substances (chapter 3). For all IT production sites throughout the world, the key question is whether MNCs and domestic firms are upgrading technological and management capacities to meet the EU standards and serve EU markets or, instead, to serve the lower standard US market.

According to the ECLAC report cited above, MNCs involved in electronics assembly in Mexico—at least along the northern border—have generally been slow to introduce innovations that reduce the toxic substances used in production and embodied in products. The problem is that there are no incentives to upgrade. On one hand, there are no national environmental standards for the IT industry in Mexico. On the other hand, Mexico-based MNCs export overwhelmingly to the US market.

In Guadalajara, we found little evidence that IT firms are retrofitting their production operations in anticipation of the new European laws. Of all the firms we interviewed, only Solectron is working toward compliance with the EU's Restriction on Hazardous Substances (RoHS). The explanation given was that the vast majority of exports from the Jalisco plants are headed for the United States, which is showing no signs of creating such a law. Some firms noted that their subsidiaries in Ireland and Hungary are already RoHS compliant because those branches of their operations export into the European market.

In the absence of national or global regulatory pressures, MNCs can—and apparently do—tailor their environmental technology and performance to particular markets. The lack of environmental upgrading in Guadalajara suggests that, at least in the short term, the IT sector will remain locked into the US market—and locked out of higher-standard markets such as the EU (CEPAL 2000).

Increasing NGO Scrutiny

FDI in the high-tech sector in Guadalajara offered some initial promise of sustainable development spill-overs. Some of the world's best companies in terms of environmental performance invested in Guadalajara. One built a state-of-the-art plant that met the world's highest standards at the time. A number banded together to develop both the supply capabilities and environmental management skills of local firms.

But the capture of significant spill-overs for sustainable development was stymied by three factors. The first was the Mexican government's failure to match FDI inflows with adequate environmental regulatory infrastructure. As a result, there was no local incentive to eco-innovate— and a significant amount of unaccounted-for hazardous IT waste is likely buried somewhere in the Mexican countryside. The second obstacle was the evolution of the IT sector into an enclave economy. With no domestic competitors, there were no direct demonstration effects on local manu-facturing firms. With a bias toward foreign contract manufacturers and few local suppliers, there were few spill-overs via backward linkages. The third obstacle was the lock-in of IT production to the US market. With its lagging environmental standards, the United States offers no incentive for Guadalajara-based firms to leapfrog to the global technology frontier.

While they appeared promising at the outset, 10 years after the start of the FDI boom it was apparent that sustainable development spill-overs were, at best, meager and at worst, missing. Interesting recent develop-ments may change this, though it is too early to tell. A coalition of global and local non-governmental organizations has recently organized to put pressure on Guadalajara's IT sector to clean up its act.

The Guadalajara-based Center for Reflection and Labor Action (CEREAL), has teamed with the UK-based Catholic Agency for Overseas Development as part of a broader CAFOD project to improve labor and environmental conditions in the global IT industry. In 2003, CEREAL provided research and input to CAFOD's report on conditions in the global electronics industry titled "How Clean Is Your Computer?" The report drew on worker interviews from electronics firms not only in Mexico but also in Europe and Asia. CEREAL and CAFOD concluded that working conditions and labor rights were poor in electronics firms— and Guadalajara was no exception.

CAFOD used its international networks and the report to "shame" foreign firms into a series of meetings that may turn out to be produc-tive. After the report was released in 2004, CEREAL hosted a handful of meetings with corporate representatives of SCI-Sanmina, Jabil Circuit, Flextronics, Solectron, and others. The first meeting was held in Guadala-jara in 2005. In attendance were members of CEREAL, representatives from CAFOD, Catholic Relief Services, CANIETI, and other NGOs. The organizers anticipate a fairly long process before companies make significant changes. Still, the willingness to dialogue is encouraging.

CEREAL points to some small successes already and is optimistic about future plans. In the first meeting, the foreign firms would not agree to the inclusion of currently employed workers in the meetings. CEREAL considered it a major achievement that workers were allowed to be present in the second meeting, which took place in 2006. At that meeting, workers from five electronics companies presented statements and opinions where they cited examples of health and safety problems such as the unsafe use of lead solder dross, as well as examples of discrimination of recruitment and dismissal without compensation. The companies formally agreed to allow workers to directly participate in future meetings and to follow up and investigate the specific complaints made by workers at the 2006 meeting.

In July of 2006, CEREAL launched a more comprehensive report titled "New Technology Workers." The report presents 73 cases of violations to Mexican laws regarding worker rights and the environment. Interestingly, many of the foreign firms agreed to comment on the report, and CEREAL published the company commentaries in the final report (CEREAL 2006). For their part, CEREAL agreed to withdraw a handful of case violations after the firms argued lack of evidence (there were originally 78, and the published report documents 73). In addition, CEREAL and CANIETI have agreed to work together to monitor the conduct of the IT firms. CANIETI has agreed to hold thematic working meetings, share information, and deal with specific cases of violations.

8

Beyond the Enclave Economy

Mexico's hopes for a dynamic IT cluster rose quickly in the mid 1990s as multinational corporations flocked to Guadalajara—and fell just as fast when the MNCs retrenched and retreated 6 years later. In the process, a domestic industry, built up during under import substitution policies from the 1940s to the 1980s, was virtually destroyed. While Guadalajara remains part of the global IT commodity chain, it will be stuck for the foreseeable future on the lower rungs of assembly and sub-assembly, have very few domestic economic linkages, and be vulnerable to highly volatile global market conditions.

The benefits to Mexico of the IT boom were meager. Rather than build linkages to local firms, FDI inflows generated an enclave: foreign firms used imported inputs to manufacture and assemble IT products for export to the United States. Moreover, the jobs created were low-skilled with little training and, absent environmental regulation, exposed workers and communities to health and ecological hazards.

Silicon and Sustainable Industrial Development

The central goal of this book has been to determine whether foreign direct investment in Mexico's IT sector in the 1990s promoted sustainable industrial development. From our findings, we aimed to draw theoretical and policy lessons about the role of FDI in development. At the outset, we determined that Mexico has been highly successful in attracting FDI, outstripping all but three developing countries in FDI inflows during the 1990s. Furthermore, a significant portion was in the IT sector clustered in Guadalajara and the state of Jalisco.

To determine whether FDI promoted sustainable industrial development, as defined by our framework, our case study sought answers to three central research questions, the first economic, the second social, and the third environmental:

- Did FDI generate knowledge spill-overs that increased the productive capacities of Mexican firms, enabling them to move up the IT value chain?
- Did FDI create sustained employment in the IT industry in Guadalajara?
- Did FDI in the IT sector transfer cleaner technologies and "best practice" environmental management to their affiliates and spur their suppliers to adopt them as well?

To answer these questions, we drew on both quantitative and qualitative analyses. Most important, we conducted more than 100 interviews with US and Mexican IT companies and industry associations, government officials, academics, journalists, workers and non-governmental organizations. Besides personal experience and insights, our interviewees gave us valuable and difficult to obtain quantitative data. As the book shows, more often than not, both methods tell the same story.

We found that FDI inflows into the IT industry in Guadalajara generated few, if any, knowledge spill-overs. Mexico's liberalization coincided with profound changes occurring in the organization of the global IT industry. Previous experts had predicted that Mexico's IT manufacturers and suppliers would be poised to form robust linkages with an influx of "global flagship" MNCs. In the event, when the foreign flagships came to Mexico (or expanded existing operations), they turned instead to foreign CMs and global, rather than national, suppliers. Made at IT corporate headquarters, the decision was not aimed specifically at Mexico but was part of flagship strategy to consolidate manufacturing in a few CMs with global reach. Nonetheless, the bias toward foreign CMs was inadvertently exacerbated by Mexican trade and industry policy. The result was a crowding out of Mexico's existing supplier base, rather than a crowding in of new investment through growing backward and forward linkages.

One earlier exception to this rule deserves note. Perhaps the most vibrant spill-overs were obtained during Mexico's first wave of IT liberalization in the 1980s rather than during the NAFTA years. Mexico exerted strong leadership when IBM sought to evade joint venture rules in the mid 1980s and establish a Guadalajara plant with 100 percent ownership. Mexico strategically leveraged the request, cutting a deal whereby IBM got 100 percent ownership in exchange for agreeing to help build local suppliers by initiating a software design and training center. All spin-off firms created by FDI were spawned by that one center (chapter 6). While China leverages every investment contract, the agreement with IBM is the only successful case of Mexico leveraging MNCs to nurture domestic firms.

On the second question, we found that the FDI initially generated a substantial number of jobs in Guadalajara's IT industry. However, these jobs were not sustained for very long—there were massive layoffs when the MNCs retrenched and relocated after 2001. Moreover, the jobs were not sustainable for individuals even during the boom. To keep within exposure limits to toxic chemicals, companies offered primarily short-term (1–12 months) contracts. Moreover, the vast majority of jobs were low-skilled and provided little or no opportunity for training and upgrading skills.

On the question of environmental performance, our findings are more mixed. On the one hand, many of the IT plants built in the 1990s were more efficient in water and energy use and less hazardous to workers in exposure to toxic materials than older plants. Lucent Technologies, for example, built a state-of-the-art facility that it wanted to use as a showcase for the developing countries. In addition, many of the foreign firms were ISO-certified for environmental management. Hewlett-Packard and other flagships required that their contract manufacturers be ISO certified as well, and even offered workshops on environmental compliance for CMs and local suppliers.

On the other hand, the CMs in Guadalajara had little to no environmental criteria for their suppliers. Moreover, the production and assembly of IT products require the use of lead, solvents and other toxic chemicals that are not well regulated in Mexico. Workers have complained about and NGOs have documented significant on-the-job hazards. In addition to occupational health, governmental regulation is poor in the area of off-site management of the toxic waste generated by

IT. Our research found that there is little accountability for hundreds of tons of waste hauled off the site by trucking companies. The best guess of environmental regulators is that it is buried somewhere in the Mexican desert.

We also found that CMs in Guadalajara, like other electronics firms along the US-Mexican border, are not producing or planning to produce to meet new, higher European environmental standards, even though their sister companies that serve the European markets were. Contrary to theories that predict that global companies will operate to global standards, our research suggests that environmental standards remain bifurcated according to the export market served by particular locales.

Why Is "Silicon Valley South" an Enclave?

Our central conclusion, in short, is that foreign direct investment in the IT sector did not promote sustainable industrial development in Mexico. Instead, it created a volatile foreign enclave in which the domestic development benefits of FDI were meager.

Why did the IT strategy fail to deliver sustainable industrial development? The root cause is the Mexican government's reliance on un-assisted FDI as the driver of development and the simultaneous downgrading of Mexico in terms of MNC global flagship strategy. In the belief that it would be sufficient to "let markets do it," Mexican policy undermined the possibility of linkage between MNCs and local firms. Policy erred both by commission and by omission. On the one hand, the PITEX program actively provided incentives to firms to import inputs, strangling demand for locally produced inputs.

In addition, Mexican macroeconomic policy kept interest rates high and the peso overvalued in order to attract and maintain foreign investment. The overvalued peso provided an additional boost to imported inputs at the expense of domestic suppliers. While foreign firms had access to internal sources of credit and global capital markets, high domestic interest rates choked off credit to Mexican firms. Coupled with largely dysfunctional financial markets, many Mexican firms simply could not obtain credit at any cost.

On the other hand, Mexico failed to design or implement policies aimed at building the technological and human capacities of local firms and workers to absorb spill-overs from MNCs. There was little investment in R&D, worker training programs, or tertiary science and engineering education. There was little in the way of a national science and technology policy. Instead, Mexico simply dismantled its previous "selective" industry policy and replaced it with "horizontal" industry policy—which meant no preferential support for domestic firms. While ostensibly providing a level playing field, the horizontal policy implicitly favored the much bigger, better financed, and linked-to-global-markets foreign firms. Even the domestic organizations established to help nurture local linkages, CANIETI and CADELEC, failed to actively promote the entrepreneurial skills and technological capacities of local firms.

In addition, Mexico did not—with the exception of the IBM deal—seek to build partnerships with MNCs to encourage knowledge transfer or local supply linkages. Indeed, the exception proves the rule. The government's partnership with IBM generated the only new firm start-ups in the whole period under study. Moreover, in concentrating solely on export markets, Mexico lost the opportunity to use FDI to diffuse IT protects in its domestic market—and to use access to a growing domestic market to leverage technology transfers by MNCs.

The failure to develop the leveraging potential of its domestic market in the early 1990s was a major blunder. Mexico's economy is second only to Brazil as the largest in Latin America. In the early 1980s, Mexico strategically leveraged knowledge transfer from IBM in exchange for market access. In the early years of the liberalization period, Mexico could have leveraged growth of and access to its domestic market to gain technology and skills upgrades to position domestic firms as suppliers to the MNCs. Such an approach could even have been embedded in NAFTA as a "carve-out." By 1996, the global IT industry had moved toward outsourcing and the huge growth of the CMs.

For their part, MNC flagships and CMs were initially attracted to Mexico as a low-wage manufacturing site in the backyard of the United States. In other words, in the immediate post-NAFTA period there was a match between a Mexico's location-specific assets and MNCs' strategic needs (Paus 2005). Yet, these assets were not sufficient to keep the MNCs in Guadalajara when the high-tech bubble burst and when China joined the WTO. The intense competitive pressures in the global IT industry make starkly clear that, to sustain FDI, countries like Mexico must invest in upgrading and expanding their location-specific assets, especially local firms' capacities to supply goods, services and niche markets, as well as a highly skilled workforce.

In short, Mexico relied solely on MNCs themselves to drive growth and upgrading in the Mexican IT sector, and matched it with little in the way of domestic effort. Drawing from Korea's successful experience in winning a secure place in the global IT industry, the analyst Linsu Kim counsels against overreliance on MNCs:

> Foreign technology transfer should not be viewed as a substitute for in-house efforts or vice versa. The two are complementary. Foreign technology transfer can provide new knowledge and serve as a catalyst for technological change, enabling firms in developing countries to make quantum jumps in technological learning. In-house efforts can, on the other hand, raise local capabilities, strengthen bargaining power in transfer negotiations and enable recipients to rapidly assimilate imported technology. (2003, p. 165)

In view of the enormous structural shift in the organization of the global IT industry—the consolidation of manufacturing in a few large CMs—a reasonable question is whether a proactive Mexican policy could really have made a difference. In a different policy environment, could local firms have evolved in concert with the MNCs into higher-level suppliers, niche producers, or even contract manufacturers?

A case study from the auto sector in India suggests that the answer is yes. Faced with liberalization after years of protection and massive FDI inflows, auto producers and suppliers in the Indian state of Tamil Nadu faced similar challenges as Guadalajara electronics firms. Indeed, a good many suffered the same fate and went out of business. However, a number not only survived but thrived, evolving into suppliers to large India-based CMs and dynamic niche producers, According to a detailed study by Meenu Tewari, the most important driver of success was government policy, especially the gradual approach of the Indian government to deregulation and liberalization. With gradual rather than "big bang" reductions in tariffs and granting of 100 percent equity stakes to foreign firms, "the pace and sequencing of the government's liberalization was highly graded and strategic" (Tewari 2003, p. 33). Today, India's automobile sector has high levels of both foreign and domestic investment and is one of the largest contributors to the country's national growth, income, and employment.

Lessons for Theory

The findings in this book underscore the need to develop more robust theories about the impact of global economic integration on developing countries, including the role of FDI in development; and about how developing countries could and should interact with the global MNCs of the 21st century to maximize local development benefits.

Since the 1980s, development theory has leaned toward the neo-classical view emphasizing the efficiency benefits of liberal trade and investment policies and the market distortions generated by selective industry policies. In recent years, however, a number of scholars have demonstrated that market failures are rife in developing countries, the world economy, and/or market economies in general. In the face of rampant market failures, government must play a proactive role in economic development.

Prevailing theories of trade and development emphasize the need to lessen the role of government in trade and industry policy. The rationale for such an approach is that domestic and international markets, when left to their own devices, will work more effectively to promote trade and development than government policy will. Indeed, reducing the role of the nation-state became a key tenet of the Washington Consensus, the prevailing "recipe" of U.S. and international economic institutions for development in the 1980s and the 1990s.

Nearly all economists agree that markets can be efficient tools for allocating scarce resources for various ends, but most also agree that there are many areas where markets break down and fail—especially in developing countries. Now that a growing body of evidence is beginning to show the mixed results of the Washington Consensus, a burgeoning number of economists are arguing that there is a need to bring the state back in to development theory and policy.

In today's globalizing world, these economists point to four key reasons why the state should still play a role in steering markets toward sustainable development—a role well beyond simply enforcing contracts, property rights, and maintaining political and economic stability. These roles center on four market failures, especially in the developing countries:

Information externalities   The private sector lacks information about opportunities to make productive investments (Lall 2005; Rodrik 2005);

Coordination externalities   Profitable new industries will not develop unless "upstream and downstream" industries are developed simultaneously (Krugman 1995; Rodrik 2005);

Imperfect competition   Economies of scale, market power, and first-mover advantages promote highly concentrated sectors that, in turn, create barriers to entry for developing-country firms. On the other hand, developing-country firms will have a better chance in the global economy (Krugman 1995) if they are oligopolies.

Environmental externalities   The environmental costs of production and consumption are not reflected in prices and lead to underproduction or overproduction of certain goods and services (Panayotou 1993).

Government policy is needed to correct these and other distortions, providing what economists call a "second best" (to markets) solution (Lall 2005; Rodrik 2005; Stiglitz 2005). When these externalities are rife in a developing country, they lead to a disincentive for the private sector to invest and innovate.

In our analysis of Mexico, these market failures were rampant. Domestic firms lacked information about the inner workings of the global IT sector. Mexico experienced coordination failures stemming from the failure to support upstream and downstream IT firms and industries to develop simultaneously. The global IT sector has many entry barriers because of imperfect competition and a generally crowded playing field. On the flip side, none of the pre-existing Mexican IT firms had the scale capabilities to break into world markets. In the longer run perhaps they could have developed dynamic efficiencies but they were clearly experiencing short run inefficiency. Finally, Mexican environmental policy is in its infancy, with the economic costs of environmental degradation estimated to be 10 percent of annual GDP (Gallagher 2004). In chapter 7 we discussed how the lack of environmental regulation in Mexico (and the United States) hinders the sustainable production of IT products.

Mexico's previous and existing policies for building domestic industry and protecting the environment—as weak as they were and are—served as second best policies to protect firms from imperfect competition and environmental externalities. Rather than creating a more level playing field, liberalization reforms exacerbated existing market failures. This has been shown to occur in economic theory. Trade theorists have shown that if trade negotiations simply liberalize rather than negotiate to correct distortions, existing distortions can worsen (Kowalcyk 2002).

Market failure, especially in relation to how it plays out in economic development processes, was studied intensively by an earlier generation of economists. In the 1950s and the 1960s, Albert O. Hirschman and other economists devoted whole books to investigating the effect of market failures on prospects for development. Unfortunately for today's quantitative economists, market failures are much harder to model than are economies with more unrealistic assumptions of perfect competition and no externalities.

For that reason, until recently, complex topics like the relationship between FDI and endogenous development have received little attention by neo-classical economists (Krugman 1995). Institutional and evolutionary economists, international political economists and sociologists have picked up the strand left vacant by Hirschman and others by conducting ex-post empirical analyses based on case study research and data gathering, rather than abstract modeling techniques, and have shed much light on these questions. We see ourselves in this latter tradition, but hope that we have teased out some insights that can bridge the two approaches.

Policy Lessons for Developing Countries

The findings in this volume have implications beyond Mexico. Developing countries across Latin America and elsewhere are seeking to increase knowledge-based assets through foreign direct investment. They see trade and investment agreements as mechanisms to attract FDI not only to increase income but also to upgrade industry and increase environmental protection.

In Mexico and Latin America, the state played a leading role in economic development from the 1950s to the 1980s. While it brought modest economic gains, the strategy was vulnerable to corruption and to generating globally uncompetitive industries and ran out of steam in the early 1980s. In response, Latin American governments swung the pendulum in the other direction—toward letting global market forces run the economy. Overall, this strategy has worked poorly for the region. Mexico was no exception. FDI crowded out domestic investment, economic growth has been slower than in the period 1950–1980, and inequality and poverty have worsened.

A re-embrace of import substitution policies is not feasible or desirable. But what is the right balance between the state and market in the 21st century, a balance that will foster sustainable industries and economic growth, raise standards of living of the poor, and protect the natural environment? Our analysis suggests six pointers to policy.

1. Treating FDI as an end in itself rather than a means to sustainable development is more likely to generate enclaves than spill-overs.

In chapter 1 we examined the gap between the theory and the reality of FDI in delivering spill-overs that promote sustainable industrial growth. The literature demonstrates unambiguously that positive FDI spill-overs are not automatic and that, in developing countries, FDI is as likely to deliver negative spill-overs—the hollowing and crowding out of domestic firms—as positive ones. Moreover, in the instances when FDI has generated positive spill-overs, it has been the result of deliberate public policies geared toward encouraging MNCs to supply them and increasing the capacities of domestic firms to absorb them. Put another away, the literature shows that it is not the quantity of FDI but the quality of the FDI–host country interface that matters. In the 1990s, Mexico's leaders, ignoring this insight, adopted an essentialist and passive development model in which the overarching objective was to increase the quantity of FDI inflows. Treating FDI as an end rather than as a means grew out of what we call the "maquila mindset"—a perception of Mexico's role as simply a low-wage, low-tax, low-tariff, export-oriented manufacturing base for North American MNCs. The strategy permeated the entire manufacturing sector. In the IT industry, there was little vision or investment in building local knowledge assets or upgrading local industrial capabilities. Without significant local knowledge assets, MNCs tend to transfer only their low-skill, low-technology operations.

A central lesson is that vision and implementing policies are essential for success in gaining positive spill-overs from FDI. Mexico's failure on both counts meant that, rather than generating spill-overs, FDI produced a disconnected enclave manufacturing economy low on innovation and heavily dependent on external suppliers and markets.

The embrace of the "hands-off" Washington Consensus is an obstacle to a strong proactive role. Even within more restrictive WTO rules that prohibit performance standards, there is much room for government leadership. Support for R&D and education is a case in point. "Latin American countries," Amsden argues (2004, p. 88), "must . . . undertake substantial policy making and institution building if it is to promote high-tech industry. The problem is not the rules of the WTO (although these generally don't help) but rather the inappropriate technology-related institutions that exist as a result of the Washington Consensus."

2. Relying on low wages alone to attract FDI leaves developing countries vulnerable to pull-out by MNCs.

Mexico's goals in attracting FDI inflows were to increase employment and income and to help balance the capital account. Low wages were central to achieving this goal. FDI inflows increased dramatically, so the strategy may appear to have been successful. However, MNCs pulled out rapidly when demand contracted in the United States and when China joined the WTO. The lesson is clear: An FDI enclave built around low wages rather than local knowledge assets and domestic markets is vulnerable to the emergence of even lower-cost producers. Moreover, as both production and transport costs fell, it became more attractive to supply North American markets from East Asian production sites. Even geographical proximity to the United States was not enough to keep MNCs in Guadalajara.

3. Garnering environmental spill-overs requires explicit attention to environmental policy.

Mexico, like other developing countries, considered IT investment "clean and green" and made environmental oversight a low priority. But significant environmental and occupational health problems are associated with IT production, problems the MNCs are still struggling to address. Moreover, the expectation that MNCs will transfer "best practice" was not borne out in Mexico. As a result, Mexican-based firms are not equipped to meet the new, high environmental standards of the European Union and will be locked out of that market. The lesson is that, like knowledge spill-overs, there is nothing automatic about environmental spill-overs from FDI. Proactive government policy is needed to generate and to capture environmental spill-overs.

4. The benefits of FDI in the IT sector are limited for late-industrializing countries.

We demonstrated in chapter 3 that there are enormous barriers to entry in the global IT industry, even at the level of third tier suppliers. The experience of Taiwan, Brazil, South Korea, and China in generating globally competitive domestic firms will be very difficult for other developing countries to emulate in the absence of strong government policy and simultaneous changes in multinational corporations' strategic interests. Without such policies, large MNCs are bound to crowd out domestic firms with global supply chains and generate few knowledge spill-overs to the local economy.

Even with proactive industry policies, however, it may not be possible for late-industrializing countries with relatively small domestic markets to enter the global IT industry except as assembly and semi-skilled manufacturing platforms for contract manufacturers. Without significant local knowledge assets, MNCs are unlikely to transfer proprietary technology. Rather, technology and skills transfer will be limited and, as we argued above, will be vulnerable to relocation to cheaper production sites.

In the short term, the employment potential of FDI in the IT sector might be substantial, though again, in the medium to long term, low-wage producers are vulnerable to footloose MNCs. Revenue potential, however, will be limited because of fierce global wage competition and the tax breaks developing countries offer to attract MNCs.

Taken together, these factors suggest that the benefits of attracting IT investment may be limited in developing countries. Other global industries, including high-tech industries, may better suit local productive capacities, domestic market potential and development objectives.

5. Trade and investment rules should expand policy space for sustainable industrial development.

The trend in trade and investment agreements, especially regional agreements such as NAFTA, is to constrict the scope for developing countries to undertake targeted industry policies. While there is still "room to move" in terms of government support to improve the overall climate for investment and innovation, the constriction of policy space impedes the potential for governments to work with FDI to nurture economic and environmental spill-overs in targeted industries.

Table 8.1 lists various policies that have been used by late-industrializing countries in the first column and indicates whether trade rules under the WTO or NAFTA permit them to be carried out. The second column reveals that the WTO has considerably more "policy space" than does NAFTA and other regional agreements. Specific to FDI, NAFTA bans joint venture requirements, technology transfer, R&D requirements, and other measures, whereas they are still permitted under the WTO. As we saw in chapter 4, China is taking full advantage of this opening, while Mexico cannot—even if it breaks the maquila mindset.

Rules on intellectual property rights, for example, make it difficult to develop comprehensive innovation policies. Investment rules outlaw the ability of developing countries to leverage concessions from foreign firms such as content requirements for local suppliers or support for local training. Investment rules also allow private foreign firms to sue national governments when new and un-anticipated (by the investing firms) social and environmental regulations cut into profits under the argument that such regulations are "tantamount to expropriation." Moreover, the macroeconomic policies needed to support contemporary trade agreements—high interest rates and tight fiscal policies—also make it more difficult for governments to design effective policy and offer credit to domestic firms.

6. Non-state actors, both NGOs and businesses, can help to make FDI work for sustainable industrial development.

In the main, governments of developed countries have worked to open developing countries to FDI by "their" MNCs, regardless of the impacts on domestic firms in host countries. Such an approach is short-sighted on

Table 8.1
Policy space for learning in WTO and NAFTA. Asterisk indicates under negotiation; X indicates no longer viable; blank indicates still viable. Source: Gallagher 2005.

Policy instrument	WTO	NAFTA
Goods trade		
Tax drawbacks		
Intellectual property		
Selective permission for patents	X	X
Short patent timelines with exceptions	X	X
Compulsory licenses		
Subsidies		
Export	X	X
R&D	*	X
Distribution	*	X
Environment	*	X
Cost of capital		
FDI		
Local content	X	X
Trade balancing	X	X
Joint ventures		X
Technology transfer		X
R&D		X
Employment of local personnel		X
Tax concessions		
Other		
Human capital		
Administrative guidance		
Movement of people	*	X
Provision of infrastructure		

a number of fronts. The investment rules in NAFTA, for example, benefited a small number of US firms, including some in the IT sector, while the slowing of GNP growth in Mexico and the hollowing out of Mexican industry reduced the export prospects of many others. Moreover, lackluster growth in manufacturing jobs in Mexico increased cross-border migration.

A better approach would be for governments of developed countries to work toward a pro-development agenda, rather than one that favors commercial interests. Our study also identifies positive roles for both business and non-governmental organizations from developed countries. For example, CAFOD and other European NGOs worked with local Mexican groups to pressure MNCs to use less environmentally destructive inputs. These efforts, and the efforts of governments, encourage "first mover" firms to adopt best practice in environmental and supply chain management and other aspects of corporate social responsibility. Lacking in the main to date, however, are sustained corporate initiatives to actively promote local industrial development, including through backward linkages.

Multinational corporations themselves can undertake initiatives to promote sustainable industrial development in the countries in which they invest—even within the context of intense global competition. On the one hand, they can increase the level of their philanthropy and make it strategic through, for example, support for education and training and partnerships with universities. On the other hand, MNCs can nurture local firms through their sourcing strategies, helping to build local firm capacities. The idea that MNCs develop local suppliers is increasingly coming to be seen as part and parcel of "corporate social responsibility."

To summarize these six points: In both theory and practice, the goal of economic integration should be to increase capacities for broad-based development at the national level. This book suggests that solely focusing on increasing levels of trade and international investment falls short of meeting larger development goals—especially in the context of sustainable industrial development. We hope that this examination of Mexico will inform discussions regarding economic integration in the Western Hemisphere, as well as the larger debates about global trade and investment rules.

Notes

Introduction

1. The word *maquila* refers to assembly plants in Mexico to which foreign materials and parts are shipped and from which finished products are delivered for sale.

2. Three have undertaken major studies, published in Spanish, of the Mexican IT sector. They are Carlos Alba Vega, Enrique Dussel, and Guillermo Woo. See the bibliography.

Chapter 1

1. For example, Motorola was paid 50.75 million pounds in 1991 to locate a mobile-phone factory in Scotland which employed 3,000 people (Haskel et al. 2001, p. 1). Ireland offers a corporate tax rate of 10 percent to all manufacturing firms (Gorg and Greenaway 2003).

2. FDI inflows may contribute to crises in the balance of payments if MNCs repatriate profits and/or increase the rate of imports faster than the rate of exports. Moreover, a high proportion of FDI inflows relative to debt may reflect not investor confidence but a high degree of country risk (Hausmann and Fernandez-Arias 2001). In this case, a large proportion of FDI in total capital inflows would suggest economic weakness, especially in domestic financial markets.

3. The massive wave of FDI inflows in the late 1990s was driven primarily by mergers and acquisitions of US by European MNCs and vice versa, as well as MNC purchase of newly privatized infrastructure assets in developing countries, especially Latin America.

4. Another determinant of FDI, generally more applicable in developed countries, is the search for new technologies.

5. NAFTA provided further impetus by requiring North American content in products produced by foreign MNCs for export to the NAFTA area.

6. In addition to Costa Rica, the six countries considered were Indonesia, Thailand, Brazil, Chile, Mexico, and Chile.

7. There are no studies of spill-overs of FDI in services in developing countries. According to UNCTAD, the presumed benefits are primarily in soft technology, that is, organizational, managerial, information processing, and other skills and knowledge, as well as more efficient inputs into primary and manufacturing industries and linkages to global markets (UNCTAD 2004, p. 124).

8. They also show that, had the authors used cross-sectional data, they would have drawn the opposite conclusion.

9. They also found that the productivity of MNC subsidiaries was higher, on average, than domestic firms.

10. Leading contemporary growth theories are the neo-classical growth model (Solow 1956) and New Endogenous Growth Theory (Romer 1994). Investment is also central to Keynesian theory, in which it drives growth through multiplier effects (Keynes).

11. For this reason, the government of Chile, in an otherwise liberal investment regime, maintains the right to exclude MNCs from domestic financial markets. The provision has never been invoked (Agosin and Mayer 2000).

12. FDI affects environmental quality by its impact on economic growth and incomes. The Environmental Kuznets Curve (EKC) posits that environmental quality first falls and then rises with per capita income. On the other hand, scale impacts of economic growth may contribute to a net decrease in local and global environmental quality, including biodiversity loss and climate change. There is little research on the scale impacts of FDI. For an explanation and critique of the EKC, see Stern 1998.

13. However, Mabey and McNally (1999) found evidence of relocation from developed to lower-standard developing host countries for resource- and pollution-intensive industries. Additionally, Cole et al. (2004) find that FDI itself contributes to the creation (or mitigation) of pollution havens, defined as increased pollution levels, depending on the degree of local government corruptibility in setting environment policy.

14. Some companies adopt the highest applicable standard, whether it is national or subnational, as the global company standard. Intel, for example, which is subject to state-level water quality regulation in each of its semiconductor production sites within the US, applies the highest standard at home to its global operations. See Leighton et al. 2002, pp. 126–130.

15. For example, 39 European companies were among the top 50 companies rated as having the best sustainability reports in 2002. Only five US companies were in the top 50 (SustainAbility 2002).

16. A landmark legal settlement in the US in April 2005 expanded liability in home countries for MNC actions in host countries. In two suits brought in California and federal courts, the California-based oil and gas company Unocal agreed to compensate Burmese villagers for complicity in human rights abuses suffered at

the hand the Burmese military, who provided security for the building of the company's natural gas pipeline in southern Burma. See Earthrights International (2005).

17. Criteria air pollutants are non-toxic air pollutants such as nitrogen oxide (NOx), sulfur oxide (SOx), sulfur dioxide (SO2), nitrogen dioxide (NO2), volatile organic compounds (VOCs), all particulates, and carbon monoxide.

18. Gallagher (2004) suggests that these policies include the reduction of sulfur levels in fuels, fuel-efficiency standards, tighter air-pollution-control standards, fuel pricing which reflects the true costs of fuel consumption, perhaps through a carbon tax, and economic incentives and disincentives.

19. For more information, see http://www.responsiblecare-us.com and http://www.icca-chem.org.

20. See http://www.ccpa.ca.

21. Some MNCs, notably in the information and communications technologies industry, are beginning to offshore R&D functions. See UNCTAD 2005.

22. See also Amsden 2001.

23. Financial institutions, in addition to community, professional, and other civil-society groups, would usefully be engaged in a partnership with MNCs to gain development benefits.

Chapter 2

1. The goal, Salinas said, was that Mexico would "export goods, not people" (Winn 1992).

2. Many foreign corporate headquarters are located in Mexico City, while their factories are elsewhere in the country. Thus, the 60 percent figure could be a data collection error.

3. The four criteria, which only cover company performance in the US, are (1) whether the firm is engaged in gambling or the production of alcohol, tobacco, or nuclear power, (2) whether the firm derives more than 2 percent of its gross revenues from the sales of military weapons, (3) strengths and weaknesses in community relations, employee relations, workforce diversity, environment and human rights, and (4) product quality and safety.

Chapter 3

1. In 2002, automobiles and other transportation equipment accounted for 26 percent of world trade. Electronics were a close second, at 19 percent.

2. The World Electronics Industry Report categorizes the electronics industry into seven segments: consumer; home appliances; data processing; telecommunications; aerospace and defense; automotive; and industry (e.g. rail transport,

power supply, etc). The IT segment corresponds to "data processing" (World Electronics Industry 2003).

3. Global flagships are sometimes called "original equipment manufacturers" or OEMs. However, in Asia, "turnkey" contract manufacturers are OEMs. To avoid confusion, we refer to the brand-name companies only as global flagships.

4. However, as design functions have become increasingly standardized, they too are being outsourced, especially to firms in East Asia. See Ernst 2004.

5. Luthje also emphasizes that total wages are highly variable because bonuses such as stock ownership and options must be oriented to customer satisfaction (Luthje 2003).

Bibliography

Aden, J., A. Kyu-Hong, and M. Rock 1999. "What is driving pollution abatement expenditure behavior of manufacturing plants in Korea?" *World Development* 27, no. 7: 1203–1214.

Agosin, M. 1996. Liberalization and the International Distribution of Foreign Direct Investment. Department of Economics, University of Chile.

Agosin, M., and R. Mayer. 2000. Foreign Investment in Developing Countries: Does It Crowd In Domestic Investment? Discussion paper 146, UNCTAD.

Aguayo Ayala, F. 2000. La estructura industrial en el modelo de economia abierta en Mexico. Working paper, Program for Science, Technology, and Development, Colegio de Mexico.

Aitken, B., and A. Harrison. 1999. "Do domestic firms benefit from foreign direct investment?" *American Economic Review* 89, no. 3: 605–618.

Alba, C. 1999. "Regional policy under NAFTA: The case of Jalisco." In *NAFTA in the New Millenium*, ed. E. Chambers and P. Smith. Center for US-Mexican Studies, University of California, San Diego.

Alcalde, A., G. Bensusán, E. de la Garza, E. Hernández Laos, T. Rendón, and C. Salas. 2000. *Trabajo y Trabajadores en el Mexico Contemporaneo*. Brígida García.

Alfaro, L. 2002. Foreign Direct Investment and Growth: Does the Sector Matter? Working paper, Harvard Business School.

Amsden, A. 2000. *The Rise of the "Rest": Challenges to the West From Late-Industrializing Economies*. Oxford University Press.

Amsden, A. 2004. "Import substitution in high-tech industries: Prebisch lives in Asia!" *CEPAL Review* 82, April: 77–91.

Amsden, A., and W. Chu. 2003. *Beyond Late Development: Taiwan's Upgrading Policies*. MIT Press.

Anderson, C. 1963. "Bankers as revolutionaries: Politics and development banking in Mexico." In *The Political Economy of Mexico*, ed. W. Glade and C. Anderson. University of Wisconsin Press.

Angel, D. and M. Rock. 2000. *Asia's Clean Revolution: Industry, Growth and the Environment*. Greenleaf Books.

Arora, A., et al. 2001. "The Indian software services industry." *Research Policy* 30: 1267–1287.

Arroyo, A. 2003. *Lessons from NAFTA: The High Cost of "Free" Trade.* Hemispheric Social Alliance.

Baer, W. 1971. "The role of government enterprises in Latin America's industrialization." In *Fiscal Policy for Industrialization and Development in Latin America*, ed. D. Geithman. University of Florida Press.

Balasubramanyam, V. 1997. "International trade in services: The case of India's computer software." *The World Economy* 20: 829–843.

BAN and SVTC (Basel Action Network and Silicon Valley Toxics Coalition). 2002. Exporting Harm: The High Tech Trashing of Asia.

Barclay, L. 2003. FDI-Facilitated Development: The Case of the Natural Gas Industry of Trinidad and Tobago. Discussion paper 2003-7, United Nations University.

Barkin, D. 1999. The Greening of Business in Mexico. UNRISD.

Barnholt, E. 2004. Remarks at Global Business and Global Poverty Conference, May 19.

Barry, D. 1995. *The Road to NAFTA. Toward a North American Community?* Westview.

Blackman, A., and X. Wu. 1998. Foreign Direct Investment in China's Power Sector: Trends, Benefits, and Barriers. Resources for the Future.

Blair, C. 1964. "Nacional financiera: Entrepeneurship in a mixed economy." In *Public Policy and Private Enterprise in Mexico*, ed. R. Vernon. Harvard University Press.

Blair, J. 1972. *Economic Concentration; Structure, Behavior and Public Policy.* Harcourt Brace Jovanovich.

Blomstrom, M., and A. Kokko. 1996. The Impact of Foreign Investment on Host Countries: A Review of the Empirical Evidence. Research working paper 1745, World Bank.

Blomstrom, M., and H. Persson. 1983. "Foreign investment and spillover efficiency in an underdeveloped economy: Evidence from the Mexican manufacturing industry." *World Development* 11: 493–501.

Blomstrom, M., and E. Wolff. 1994. "Multinational corporations and productivity convergence in Mexico." In *Convergence of Productivity*, ed. W. Baumol et al. Oxford University Press.

Borensztein, E. 1998. "How does foreign direct investment affect economic growth?" *Journal of International Economics* 45: 115–135.

Bosworth, B., and S. Collins. 1999. *Capital Flows to Developing Economies: Implications for Savings and Investment.* Brookings Institution.

Brannon, J. 1994. "Generating and sustaining backward linkages between maquiladoras and local suppliers in northern Mexico." *World Development* 22, no. 12: 1933–1945.

Bruton, H. 1998. "A reconsideration of import substitution." *Journal of Economic Literature* 36, no. 2: 903–936.

Business for Social Responsibility. Undated. Monitoring of Global Supply Chain Practices. Issue brief.

CADELEC (Cadena productiva de la electronica). 2004. Retrieved from www.cadalec.com.mx.

CAFOD (Catholic Action for Overseas Development). 2004. "Clean up your computer: Working conditions in the electronics sector." Retrieved from http://www.cafod.org.uk.

CANIETI (Camara Nacional de la industria electronica). 2003. "Antecedentes." Retrieved from http://www.canieti.org.

Caves, R. 1974. "Multinational firms, competition and productivity in host-country markets." *Economica* 41, no. 162: 176–193.

CEPAL (Economic Commission of Latin America and the Caribbean). 2000. Foreign Investment in Latin America and the Caribbean.

CEREAL (Centro de Refelexion y Accion Laboral). 2006. New Technology Workers: Report on Working Conditions in the Mexican Electronics Industry.

Chandler, A. 2001. *Inventing the Electronic Century: The Epic Story of the Consumer Electronics and Computer Industries*. Free Press.

Chang, H.-J. 2002. *Kicking Away the Ladder: Development Strategy in Historical Perspective*. Anthem.

Christensen, C. 1997. *The Innovator's Dilemma: When New Technologies Cause Great Firms to Fail*. Harvard Business School Press.

Chudnovsky, D., A. Lopez, and G. Rossi. 2004. Foreign Direct Investment Spillovers and the Absorption Capabilities of Domestic Firms in the Argentine Manufacturing Sector, 1992–2001. Mercosur Economic Research Network, Universidad de San Andres.

Cimoli, M., ed. 2000. *Developing Innovation Systems: Mexico in a Global Context*. Continuum.

Cohen, W., and D. Levinthal. 1990. "Absorptive capacity: A new perspective on earning and innovation." *Administrative Science Quarterly* 35: 128–152.

Contreras, J. 2006. "What China threat?" *Newsweek*, May 15: 36–37.

Courier (ACP–European Union). 2003. "Sustainable industrial development: Alleviating poverty by fostering productivity growth. An Interview with Maurizio Carbone." January-February.

CSIS (Center for Strategic International Studies). 2003. Economic Competitiveness in Mexico.

Dasgupta, S., D. Wheeler, and H. Hettige. 1997. What Improves Environmental Performance? Evidence from Mexican Industry. Development Research Group, World Bank.

Dasgupta, S., H. Hettige, and D. Wheeler. 2000. "What improves environmental compliance? Evidence from Mexican Industry." *Journal of Environmental Economics and Management* 38: 39–66.

Dedrick, J., and K. Kraemer. 2002a. "Enter the dragon: China's computer industry." *Perspectives*, February: 28–36.

Dedrick, J., and K. Kraemer. 2002b. Globalization of the personal computer industry: Trends and implications. Center for Research on Information Technology and Organizations, University of California, Irvine.

Dedrick, J., et al. 2001. Liberalization of Mexico's Computer Sector. Center for Research on Information Technology and Organization, University of California, Irvine.

Dominquez-Villalabos, L. 2000. "Environmental performance in the Mexican chemical fibres industry in the context of an open market." In *Industry and Environment in Latin America*, ed. R. Jenkins. Routledge.

Dornbusch, R., and A. Werner. 1994. "Mexico: Stabilization, reform, and no growth." *Brookings Papers on Economic Activity* 1: 253–315.

Dussel, E. 1999. "Reflexiones sobre conceptos y experiencias internacionales de industrializacion regional." In *Dinamica Regional y Competitividad Industrial*, ed. C. Ruiz Durán and E. Dussel. Editorial JUS.

Dussel, E. 2003. "Industrial policy, regional trends, and structural change in Mexico's manufacturing sector." In *Confronting Development*, ed. K. Middlebrook and E. Zepeda. Stanford University Press.

Dussel, E. 2004. "Gastos en investigacion y desarrollo e inversion extranjera directa: Un estudio a nivel de clases economicas del sector manufacturero Mexicoano, 1994 to 2000." In *Nuevos Caminos Para el Desarollo Sustentable en Mexico*, ed. A. Nadal. Siglo XXI, Mexico.

Dussel, E. 2005. *Economic Opportunities and Challenges Posed by China for Mexico and Central America*. German Development Institute.

Dussel, E., A. Partida, and G. Woo. 2003. La industria Electronica en Mexico: Problematica, Perspectivas y Propuestas. Universidad de Guadalajara.

ECLAC (Economic Commission on Latin America and the Caribbean). 2003. La industria maquiladora electronica en la frontera norte de Mexico y el medio ambiente. Report LC/MEX/L.585.

Edwards, S., and M. Savastano. 1998. *The Morning After: The Mexican Peso in the Aftermath of the 1994 Currency Crisis*. National Bureau of Economic Research.

Ernst, D. 2000. The Economics of Electronics Industry: Competitive Dynamics and Industrial Organization. Working paper 7, East-West Center.

Ernst, D. 2003. How Sustainable Are Benefits from Global Production Networks? Malaysia's Upgrading Prospects in the Electronics Industry. East-West Center.

Ernst, D., and L. Kim. 2001. Global Production Networks, Knowledge Diffusion, and Local Capability Formation: A Conceptual Framework. Working paper 19, East-West Center.

Eskeland, G. S., and A. Harrison. 1997. *Moving to Greener Pastures? Multinationals and the Pollution Haven Hypothesis.* World Bank.

Evans, P. 1995. *Embedded Autonomy: States and Industrial Transformation.* Princeton University Press.

Federal Reserve Bank of Dallas. 2003. China: Awakening Giant.

Findlay, R. 1978. "Relative backwardness, direct foreign investment, and the transfer of technology: A simple dynamic model." *Quarterly Journal of Economics* 92: 1–16.

Flextronics. 2003. Environmental Health and Safety Policy. Retrieved from http://www.flextronics.com.

Foran, T. 2001. Corporate Social Responsibility at Nine Multinational Electronics Firms in Thailand. Report to California Global Corporate Accountability Project, Nautilus Institute.

Gallagher, K. 2002. "Industrial pollution in Mexico: Did NAFTA matter?" In *FTAA and the Environment*, ed. C. Deere and D. Esty. MIT Press.

Gallagher, K. 2003. "The CEC and environmental quality: Assessing the Mexican experience." In *Greening NAFTA*, ed. J. Knox and D. Markel. Stanford University Press.

Gallagher, K. 2004. *Free Trade and the Environment: Mexico, NAFTA, and Beyond.* Stanford University Press.

Gallagher, K., ed. 2005. *Putting Development First: The Importance of Policy Space in the WTO and IFIs.* Zed.

Gallagher, K. S. 2006. *China Shifts Gears: Automobiles, Oil, Pollution, Development.* MIT Press.

Garcia-Johnson, R. 2000. *Exporting Environmentalism: U.S. Multinational Chemical Corporations in Brazil and Mexico.* MIT Press.

GATT (General Agreement on Tariffs and Trade). 1993. *Trade Policy Review: Mexico 1993.* GATT Publications.

Gentry, B., ed. 1998. *Private Capital Flows and the Environment: Lessons from Latin America.* Elgar.

Gentry, B., and L. Fernandez. 1998. "Mexican steel." In *Private Capital Flows and the Environment*, ed. B. Gentry. Elgar.

Gerber, J., and J. Carrillo. 2003. "Las maquiladoras de Baja California son competitivas?" *Comercio Exterior* 53, no. 3: 284–293.

Gereffi, G., and T. Sturgeon. 2004. Globalization, Employment, and Economic Development. Sloan Workshop Series in Industry Studies.

Gerschenkron, A. 1962. *Economic Backwardness in Historical Perspective.* Harvard University Press.

Gonzalez, G. 1999. "To what extent does investment equal growth?" Interpress News Service, August 20.

Gorg, H., and D. Greenaway. 2004. "Much Ado About Nothing? Do Domestic Firms Really Benefit from Foreign Direct Investment?" *World Bank Research Observer* 19, no. 2: 171–197.

Gori, G. 2002. "Fears about Microsoft return in Mexico." *New York Times*, April 26.

Greenpeace Mexico. 2005. "Exige Greenpeace a industria electrónica eliminar compuestos tóxicos de sus productos." Retrieved from http://www.greenpeace. org.

Grossman, G. 1992. *Imperfect Competition and International Trade.* MIT Press.

Hansen, M. W. 1999. Cross border environmental management in transnational corporations. An analytical framework. Occasional paper 5, UNCTAD/CBS Project.

Hanson, G. H. 2003. What Has Happened to Wages in Mexico since NAFTA? Implications for Hemispheric Free Trade. Working paper 9563, National Bureau of Economic Research.

Hettige, H., M. Huq, S. Pargal, and D. Wheeler. 1996. "Determinants of pollution abatement in developing countries: Evidence from South and Southeast Asia." *World Development* 24, no. 12: 1891–1904.

Hirschman, A. 1958. *The Strategy of Economic Development.* Yale University Press.

INEGI (Instituto Nacional de Estadística Geografía e Informática). 2000. Sistema de Cuentas Economicas y Ecologicas de Mexico, 1993–1999.

INEGI. 2003. Censos Economicos. Retrieved from www.inegi.gob.mx.

ISIS Asset Management Inc. 2004. The ICT Sector: The Management of Social and Environmental Issues in Supply and Disposal Chains.

ISOWorld. 2003. Number of ISO14001 Certifications in the World. Retrieved from www.ecology.or.jp.

Jabil. 1999. "Jabil announces quality certifications." Retrieved from http://www. jabil.com.

Jacobs, S. 2002. "E-Mexico revisited." *Business Mexico* 12, no. 3: 55.

Jaffe, A., S. Peterson, and Portney, P. 1995. "Environmental regulation and the competitiveness of U.S. manufacturing: What does the evidence tell us?" *Journal of Economic Literature* 33, March: 132–163.

Kehoe, T. J. 1995. "A review of Mexico's trade policy from 1982–1994." *The World Economy* 18: 135–151.

Keller, W., and S. R. Yeaple. 2003. Multinational Enterprises, International Trade, and Productivity Growth: Firm Level Evidence from the United States. Working paper 9504, National Bureau of Economic Research.

Kim, L. 2003. "The dynamics of technology development: Lessons from the Korean experience." In *Competitiveness, FDI and Technological Activity in East Asia,* ed. S. Lall and S. Urata. Elgar.

Koch, C. 2001. "Yank your chain." Retrieved from Darwinmag.com.

Kokko, A., R. Tansini, and M. Zejan. 1996. "Local technological capability and productivity spillovers from FDI in the Uruguayan manufacturing sector." *Journal of Development Studies* 34: 602–611.

Kowalcyk, C. 2002. "Reforming tariffs and subsidies in international trade." *Pacific Economic Review* 7, no. 2: 552–559.

Kraemer, K., and J. Dedrick. 2002. Globalization of the Personal Computer Industry: Trends and Implications. Center for Research on Information Technology and Organizations, University of California, Irvine.

Krugman, P. 1995. *Development, Geography, and Economic Theory*. MIT Press.

Krugman, P. 1998. Firesale FDI. Working paper, Massachusetts Institute of Technology.

Kuczynski, P.-P., and J. Williamson. 2003. *After the Washington Consensus: Restarting Growth and Reform in Latin America*. Institute for International Economics.

Kuehr, R., and E. Williams, eds. 2003. *Computers and the Environment, Understanding and Managing Their Impacts*. Kluwer.

Kumar, N. 2002. Foreign Direct Investment, Externalities and Economic Growth: Some Empirical Explorations for Low-Income Countries. Research Seminar, United Nations University Institute for New Technologies, Maastricht.

Kumar, N., and K. Joseph. 2004. The National Innovation System That Made India's IT Success Possible: Are There Any Lessons for New ASEAN Member Countries? Research and Information System for the Non-Aligned and Other Developing Countries, New Dehli.

Lagos, G., and P. Velasco. 1999. "Environmental policies and practices in Chilean mining." In *Mining and the Environment*, ed. A. Warhurst. International Development Research Centre.

Lall, S. 2005. "Rethinking industrial strategy: The role of the state in the face of globalization." In *Putting Development First*, ed. K. Gallagher. Zed.

Lall, S., and S. Urata, eds. 2004. *Competitiveness, FDI and Technological Activity in East Asia*. Elgar.

Larrain, F., L. Lopez-Calva, and A. Rodriguez-Clare. 2000. Intel: A Case Study of Foreign Direct Investment in Central America. Working paper 58, Center for International Development, Harvard University.

Leighton, M., N. Roht-Arriaza, and L. Zarsky. 2002. *Beyond Good Deeds: Case Studies and a New Policy Agenda for Global Corporate Accountability*. Natural Heritage Institute.

Lim, E.-G. 2001. Determinants of and the Relation between Foreign Direct Investment and Growth: A Summary of the Recent Literature. Working paper 01/175, International Monetary Fund.

List, F. 1885. *The National System of Political Economy*. Longman.

Lowe, N., and M. Kenney. 1999. "Foreign investment and the global geography of production: Why the Mexican consumer electronics industry failed." *World Development* 27, no. 8: 1427–1443.

Luhnow, D. 2004. "As jobs move east, plants in Mexico retool to compete." *Wall Street Journal*, March 5.

Lustig, N. 1992. *Mexico: The Remaking of an Economy*. Brookings Institution.

Luthje, B. 2003. IT and the Changing Social Division of Labor: The Case of Electronics Contract Manufacturing. Working paper, Institute of Social Research, University of Frankfurt.

Luxner, L. 2000. "Intel factory leads Costa Rica's high tech boom." Retrieved from www.luxner.com.

Maquilaportal. 2005. "New director for Jabil Circuit." Retrieved from http://www.maquilaportal.com.

Marin, A., and M. Bell. 2003. Technology Spillovers from Foreign Direct Investment (FDI): An Exploration of the Active Role of MNC Subsidiaries in the Case of Argentina in the 1990s. Discussion paper, Science and Technology Policy Research Unit, University of Sussex.

Markoff, J. 2003. "Innovation at Hewlett tries to evade the ax." *New York Times*, May 5.

Mayer, F. 1998. *Interpreting NAFTA: The Science and Art of Political Analysis*. Columbia University Press.

Mazurek, J. 1999. *Making Microchips: Policy, Globalization and Economic Restructuring in the Semiconductor Industry*. MIT Press.

Mercado, A. 2000. "Environmental assessment of the Mexican steel industry." In *Industry and the Environment in Latin America*, ed. R. Jenkins. Routledge.

Mexico. 1989. Plan Nacional de Desarrollo, 1989–1994. Secretaria de Programacion y Presupuesto.

Mohin, T. 2005. The IC industry: Past environmental progress, challenges and opportunities for the future. Presented at SEMICON Semiconductor Market Forum, Shanghai.

Montoy, A. 2001. "Interview with Mr. Alejandro Gomez Montoy, Selectron de Mexico." *Far Eastern Economic Review*.

Moran, T. H. 1998. *Foreign Direct Investment and Development: The New Policy Agenda for Developing Countries and Economies in Transition*. Institute for International Economics.

Moreno, J., and J. Ros. 1994. "Market reform and the changing role of the state in Mexico: A historical perspective." In *Empirical Studies: The State, Markets and Development*, ed. A. Dutt et al. Elgar.

MOST (Ministry of Science and Technology, People's Republic of China). 2006. National High Tech R&D Program (863 Program).

Nadal, A. 2003. "Macroeconomic challenges for Mexico." In *Confronting Development*, ed. K. Middlebrook, and E. Zepeda. Stanford University Press.

Nelson, R. 2001. *The Sources of Economic Growth*. Harvard University Press.

Nunnenkamp, P., and Spatz, J. 2002. "Determinants of FDI in developing countries: Has globalization changed the rules of the game?" *Transnational Corporations* 11, no. 2: 1–34.

Occupational Knowledge International. 2005. Environmental Certification Program for Lead Battery Manufacturing. Retrieved from www.okinternational.org.

Ochoa, M. 2003. Interview by the authors, October 27.

OECD (Organization for Economic Cooperation and Development). 1998. Environmental Performance Review for Mexico.

OECD. 2002. The Environmental Benefits of FDI.

OECD. 2002. Foreign Direct Investment for Development: Maximising Benefits, Minimising Costs.

Otero, G. 1996. "Neoliberal reform and politics in Mexico." In *Neo-Liberalism Revisited*, ed. G. Otero. Westview.

Palacios, J. 2001. Production Networks and Industrial Clustering in Developing Regions: Electronics Manufacturing in Guadalajara, Mexico. UDG Press.

Panayotou, T. 1993. *Green Markets: The Economics of Sustainable Development*. ICS Press.

Partida, R. 2004. Efectos de los Tratados de Libre Comercio suscritos con otros paises sobre los salarios y empleo (NAFTA, Union Europea). Working paper, Universidad de Guadalajara.

Partida, R., and P. Moreno. 2003. "Redes de vinculacion de la Universidad de Guadalajara con la industria electronica de la Zona Metropolitana." In *La Industria Electronica en Mexico*, ed. E. Dussel et al. Universidad de Guadalajara.

Pastor, M. 1998. "Pesos, policies, predictions." In *The Post-NAFTA Political Economy*, ed. C. Wise. Johns Hopkins University Press.

Paus, E. 2005. *Foreign Investment, Development, and Globalization: Can Costa Rica Become Ireland?* Palgrave.

Peres, W. 1990. *Foreign Direct Investment and Industrial Development in Mexico*. OECD.

Pohl, O. 2006. European environmental rules propel change in the U.S. *New York Times*, September 23.

Rainforest Alliance. 2000. "News from the front: Banana eco-labeling program is world's largest." Retrieved from www.rainforestalliance.org.

Rasiah, R. 2003. "Industrial technology transition in Malaysia." In *Competitiveness, FDI and Technological Change in East Asia*, ed. S. Lall and S. Urita. Elgar.

Raymond Communications. 2003. Backgrounder: Electronic Waste. Retrieved from http://www.raymond.com.

Razin, A., E. Sadka, and C. Yuen. 1999. Excessive FDI Flows under Asymmetric Information. Working paper 7400, National Bureau of Economic Research.

Reynolds, C. 1970. *The Mexican Economy*. Yale University Press.

Rivera Vargas, M. 2002. *Technology Transfer via the University-Industry Relationship: The Case of Foreign High Technology Electronics Industry in Mexico's Silicon Valley*. Routledge.

Roberts, T. 2004. "How to make computers greener." Retrieved from http://news.bbc.co.uk.

Rock, M., and D. Angel. 2005. *Industrial Transformation in the Developing World*. Oxford University Press.

Rodrik, D. 1999. *The New Global Economy and Developing Countries: Making Openness Work*. Overseas Development Council.

Rodrik, D. 2005. Industrial Policy for the 21st Century. UNIDO.

Romm, J. 1999. The Internet Economy and Global Warming. Retrieved from www.cool-companies.org.

Romo Murillo, D. 2002. "Derramas tecnologicas de la inversion extranhera en la industria Mexicana." *Comercio Exterior* 53, no. 3: 230–243.

Rosenthal, E. 2002. "Conflicts over transnational oil and gas development off Sakhalin Island in the Russian Far East: A David and Goliath tale." In *Human Rights and the Environment*, ed. L. Zarsky. Earthscan.

Ruud, A. 2002. "Environmental management of transnational corporations in India: Are TNCs creating islands of environmental excellence in a sea of dirt?" *Business Strategy and the Environment* 11, no. 2: 103–118.

Saagi, K. 2002. Trade, Foreign Direct Investment, and International Technology Transfer: A Survey. Background paper for World Bank's Microfoundations of International Technology Diffusion project, Department of Economics, Southern Methodist University.

Sachs, J., A. Tornell, and A. Velasco. 1995. "The collapse of the Mexican peso." *Journal of InterAmerican Studies and World Affairs* 37, no. 2: 57–81.

Salas, C. 2002. Highlights of current labor market conditions in Mexico. La Red de Investigadores y Sindicalistas para Estudios Laborales. Global Policy Network.

Salas, C., and E. Zepeda. 2003. "Employment and wages: Enduring the costs of liberalization." In *Confronting Development*, ed. K. Middlebrook and E. Zepeda. Stanford University Press.

Sawada, N. 2004. Technology Spillovers and Welfare through Foreign Direct Investment in Developing Countries: An Oligopoly Approach. Presented at Mid-West International Economics Meetings.

Schatan, C. 2002. "Mexico's manufacturing exports and the environment under NAFTA." In *The Environmental Effects of Free Trade*. North American Commission for Environmental Cooperation.

Semiconductor Industry Association. 2004. SIA Seeks Proposals for Worker Health Study. Press release, August 19.

SEPROE (Secretary of Economic Promotion, Jalisco). 2001. Jalisco y sus Sectores Estrategicos.

Shaiken, H. 1990. Mexico in the Global Economy: High Technology and Work Organization in Export Industries. Center for U.S.-Mexican Studies, University of California, San Diego.

Shang, S., W. Tu, W. Yan, and L. Yang. 2003. Environmental and Social Aspects of Taiwanese and U.S. Companies in the Hsinchu Science-Based Industrial Park. Report to California Global Corporate Accountability Project, Nautilus Institute.

Sherman, R. 2004. Electronics Manufacturing in China and Mexico. Presented at Mexico and China Conference, Chicago.

Smarzynska, B. 2003. Does Foreign Direct Investment Increase the Productivity of Domestic Firms? In Search of Spillovers through Backward Linkages. William Davidson Working paper 548, University of Michigan Business School.

Soreide, T. 2001. FDI and industrialization: why technology structure and new industrial structures may accelerate economic development. Working paper 2001:3, Christian Michaelsen Institute.

Spooner, J. G. 2005. "The nouveau Lenovo wants to shake up the PC market's status quo." Retrieved from http://www.zdnet.com.

Stiglitz, J. 2005. "Development policies in a world of globalization." In *Putting Development First*, ed. K. Gallagher. Zed.

Stromberg, P. 2002. The Mexican Maquila Industry and the Environment: An Overview of the Issues. Working paper, Naciones Unidas CEPAL/ECLAC.

Sturgeon, T. 2002. "Modular production networks: A new American model of industrial organization." *Industrial and Corporate Change* 11, no. 3: 451–496.

TCPS (Texas Center for Policy Studies). 2004. The Generation and Management of Hazardous Wastes and Transboundary Hazardous Waste Shipments between México, Canada and the United States since NAFTA: A 2004 Update. Retrieved from http://www.texascenter.org.

Ten Kate, A. 1992. "Trade liberalization and economic stabilization in Mexico: Lessons of experience." *World Development* 20, no. 5: 659–672.

Tewari, M. 2003. Foreign Direct Investment and the Transformation of Tamil Nadu's Automotive Supply Base. Retrieved from http://www.globalvaluechains. org.

UNCTAD (United Nations Commission for Trade and Development). 1996. World Investment Report: Investment, Trade and International Policy Arrangements.

UNCTAD. 2000. World Investment Report, Crossborder Mergers and Acquisitions and Development.

UNCTAD. 2002. Foreign Direct Investment Statistics.

UNCTAD. 2004. World Investment Report, The Shift Towards Services.

UNCTC (United Nations Centre on Transnational Corporations). 1992. Foreign Direct Investment and Industrial Restructuring in Mexico.

UNDP (United Nations Development Programme). 2001. *Human Development Report.* Oxford University Press.

UNIDO (United Nations Industrial Development Organization). 2002. Competition Through Innovation and Learning. Industrial Development Report 2002/2003.

UNIDO. 2005. Monitor de la Manufactura Mexicana.

United Nations. 2006. Commodity Trade Statistics (COMTRADE).

USDOC (U.S. Department of Commerce). 2006. Technology Transfer to China. Retrieved from www.bis.doc.gov.

Whiting, V. 1992. *The Political Economy of Foreign Investment in Mexico: Nationalism, Liberalism, and Constraints on Choice.* Johns Hopkins University Press.

Wilson, P. 1992. *Exports and Local Development: Mexico's New Maquiladoras.* University of Texas Press.

Winn, P. 1992. *Americas: The Changing Face of Latin America and the Caribbean.* University of California Press.

Wise, C., ed. 1998. *The Post-NAFTA Political Economy: Mexico and the Western Hemisphere.* Pennsylvania State University Press.

Wisner, P., and M. Epstein. 2003. The NAFTA Impact on Environmental Responsiveness and Performance in Mexican Industry. Unpublished paper.

Wong, P. K. 2003. "From using to creating technology: The evolution of Singapore's national innovation system and the changing role of public policy." In *Competitiveness, FDI and Technological Activity in East Asia,* ed. S. Lall and S. Urata. Elgar.

Woo, G. 2001. "Hacia la integracion de pequenas empresas en la industria electronica de Jalisco: Dos casos de estudio." In *Claroscuros,* ed. E. Dussel. Editorial Jus.

World Bank. 1998. Mexico: The Guadalajara Environmental Management Pilot.

World Bank. 2005. *World Development Indicators.*

World Bank. 2006. *World Development Indicators.*

Zarsky, L. 2002. "Stuck in the mud? Nation states, globalization, and the environment." In *The Earthscan Reader on International Trade and Sustainable Development,* ed. K. Gallagher and J. Werksman. Earthscan.

Zarsky, L., ed. 2005. *International Investment for Sustainable Development: Balancing Rights and Rewards.* Earthscan.

Urban and Industrial Environments: The Series

Maureen Smith, *The U.S. Paper Industry and Sustainable Production: An Argument for Restructuring*

Keith Pezzoli, *Human Settlements and Planning for Ecological Sustainability: The Case of Mexico City*

Sarah Hammond Creighton, *Greening the Ivory Tower: Improving the Environmental Track Record of Universities, Colleges, and Other Institutions*

Jan Mazurek, *Making Microchips: Policy, Globalization, and Economic Restructuring in the Semiconductor Industry*

William A. Shutkin, *The Land That Could Be: Environmentalism and Democracy in the Twenty-First Century*

Richard Hofrichter, ed., *Reclaiming the Environmental Debate: The Politics of Health in a Toxic Culture*

Robert Gottlieb, *Environmentalism Unbound: Exploring New Pathways for Change*

Kenneth Geiser, *Materials Matter: Toward a Sustainable Materials Policy*

Thomas D. Beamish, *Silent Spill: The Organization of an Industrial Crisis*

Matthew Gandy, *Concrete and Clay: Reworking Nature in New York City*

David Naguib Pellow, *Garbage Wars: The Struggle for Environmental Justice in Chicago*

Julian Agyeman, Robert D. Bullard, and Bob Evans, eds., *Just Sustainabilities: Development in an Unequal World*

Barbara L. Allen, *Uneasy Alchemy: Citizens and Experts in Louisiana's Chemical Corridor Disputes*

Dara O'Rourke, *Community-Driven Regulation: Balancing Development and the Environment in Vietnam*

Brian K. Obach, *Labor and the Environmental Movement: The Quest for Common Ground*

Peggy F. Barlett and Geoffrey W. Chase, eds., *Sustainability on Campus: Stories and Strategies for Change*

Steve Lerner, *Diamond: A Struggle for Environmental Justice in Louisiana's Chemical Corridor*

Jason Corburn, *Street Science: Community Knowledge and Environmental Health Justice*

Peggy F. Barlett, ed., *Urban Place: Reconnecting with the Natural World*

David Naguib Pellow and Robert J. Brulle, eds., *Power, Justice, and the Environment: A Critical Appraisal of the Environmental Justice Movement*

Eran Ben-Joseph, *The Code of the City: Standards and the Hidden Language of Place Making*

Nancy J. Myers and Carolyn Raffensperger, eds., *Precautionary Tools for Reshaping Environmental Policy*

Kelly Sims Gallagher, *China Shifts Gears: Automakers, Oil, Pollution, and Development*

Kerry H. Whiteside, *Precautionary Politics: Principle and Practice in Confronting Environmental Risk*

Ronald Sandler and Phaedra C. Pezzullo, eds., *Environmental Justice and Environmentalism: The Social Justice Challenge to the Environmental Movement*

Julie Sze, *Noxious New York: The Racial Politics of Urban Health and Environmental Justice*

Robert D. Bullard, ed., *Growing Smarter: Achieving Livable Communities, Environmental Justice, and Regional Equity*

Ann Rappaport and Sarah Hammond Creighton, *Degrees That Matter: Climate Change and the University*

Michael Egan, *Barry Commoner and the Science of Survival: The Remaking of American Environmentalism*

David J. Hess, *Alternative Pathways in Science and Industry: Activism, Innovation, and the Environment in an Era of Globalization*

Peter F. Cannavò, *The Working Landscape: Founding, Preservation, and the Politics of Place*

Paul Stanton Kibel, ed., *Rivertown: Rethinking Urban Rivers*

Kevin P. Gallagher and Lyuba Zarsky, *The Enclave Economy: Foreign Investment and Sustainable Development in Mexico's Silicon Valley*

Index

Acer group, 78, 104
Apertura policy, 47, 49
Apparel industry, 37, 45, 54, 55
Appliance industry, 72, 136
Assembly, 45, 50, 54, 64, 65, 73, 101,
    109, 133, 135, 162, 166
Audio-visual industry, 72, 126, 129
Auto and auto-parts industries, 24,
    34, 39, 50, 54, 55, 60, 72, 136,
    182, 183

Biotech industry, 119
Burroughs, 128, 141, 145

CADELEC, 131, 132, 150, 154, 180
CAFOD, 170, 175, 191
CANIETI, 131, 154, 175, 176
Carcinogens, 90, 91, 116
Cathode-ray tubes, 92, 93
Chemical hazards, 89–95, 115–117,
    162, 163, 171
Chemical industry, 35, 47, 57, 69
Circuit boards, 49, 90, 92, 129, 130,
    149, 162
Collaboration, 20, 29, 36, 37, 40, 60,
    118
Components industry, 72, 74, 77–79,
    90, 91, 111, 144–146
Computers, 72, 81–89, 92, 93,
    102–104, 108, 111, 113, 114, 122,
    125, 128–131, 141, 149
Contract manufacturers, 72,
    75–77, 81–84, 89, 95, 96, 100–104,

109–111, 121, 129, 130, 135–156,
    166, 170–172, 178–183
Credit markets, 14, 40, 46, 47, 51,
    53, 57–59, 106, 108, 113, 123,
    153–156, 180

Debt
    domestic, 40, 58, 59, 123
    foreign, 46–53, 123, 125, 155, 156
Deficit, 46–49, 123
Dell Computer, 81, 114, 172
Demonstration effect, 21, 36, 51, 60,
    172, 175
Development theory, 183–185

Economic development, 13, 14, 20,
    24, 25, 36, 37, 40–45, 48–64, 78,
    79, 100, 110, 118, 131, 134–140,
    180–191
Education, 39–41, 51, 60, 101,
    105–110, 113, 114, 118, 119, 131,
    146–155, 179, 181, 187
Electrónica Pantera, 128, 141,
    144–146
Electronics industry, 23, 24, 39, 49,
    54, 55, 72–81, 84–87, 94–119, 122,
    126–133, 139–160, 172, 188–191
e-Mexico program, 152
Energy technologies, 32, 41, 89, 93, 95
Entrepreneurs, 46–52, 99, 125, 126,
    144–149, 156
Environmental initiatives, voluntary,
    68, 69, 94, 96

Environmental management, 14, 18, 28–35, 40–42, 63–69, 88–97, 115–119, 159–180, 185, 188
Equipment industries, 39, 49, 50, 54, 55
Exchange rate, 13, 43–48, 52, 55–59, 106, 125, 156, 156, 180
Exports, 41, 56, 57, 61, 63–67, 100–104, 109, 110, 113–115, 119, 122–126, 131–133, 156, 156

Flextronics, 76, 95, 101, 129, 136, 141, 143, 171, 172, 175
Food and beverage industries, 57, 62
Foreign direct investment, 13–19, 28, 40–57, 99–115, 119, 121–136, 139–144, 152–156, 166, 177, 178, 182–191
Fox, Vicente, 52, 53, 126, 151, 152
Free-market ideology, 13, 27, 35–37, 40, 125, 126, 180, 187
Furniture industry, 60, 65

General Motors, 60, 69, 136
Global production network, 72–83, 88, 100, 101, 109, 112, 129, 130, 134, 140–152, 166–170, 180–183
Greenfields, 14, 29, 43, 54
Greenhouse gases, 94, 96, 166–170
Guadalajara, 124, 125, 128–177, 182

Hewlett-Packard, 69, 72, 83, 84, 93, 111–114, 123, 129, 135, 141, 144, 163, 168–172, 179
Human capital, 21, 101, 102, 146–150, 181

IBM, 69–75, 81–83, 90, 111, 114, 122–125, 128, 129, 135, 141–144, 148, 151, 163, 169, 172, 173, 179, 181
Imports, 16, 17, 44–48, 55–57, 61, 103–106, 131, 143, 144, 156, 185
Import-substituting industrialization, 44–48, 105, 122, 123, 129, 177
Industrial parks, 108, 112, 113, 116, 131

Inflation, 46–48, 56–59
Innovation, 17, 19, 27–29, 53, 61, 62, 72–75, 78–80, 100, 102, 110, 114, 154, 187
Institutionalism, 105–108, 111, 118, 122, 124, 154
Intel, 15, 19, 83, 93, 94, 129, 148, 149
Intellectual property, 30, 36, 50, 51, 99, 189
Interest rates, 26, 43, 46, 56–59, 113, 123, 136, 139, 156, 156, 180
International Monetary Fund, 47, 49, 52
ISO certification, 143, 163, 164, 168–173, 179

Jabil Circuit, 76, 129, 135, 136, 141, 143, 171, 175
Jalisco, 126–141, 148, 154–172, 177
Joint ventures, 104, 111–115, 141, 145, 154, 179

Knowledge, 14, 19–26, 30, 35–38, 41, 50–53, 59, 61, 75, 80, 81, 99–102, 140–142, 178–181, 186–190

Labor unions, 48–51, 77, 131, 175, 177
Lead, exposure to, 91, 92, 95, 115–117, 160, 162, 170, 171, 176, 179
Leapfrogging, 28, 35, 36, 102, 105, 110, 173
Liability, environmental, 31, 32
Liberalization
"big bang," 122–128
as economic policy, 41, 78, 185
and global production, 78, 80, 140
of Mexico's policies, 43, 44, 48, 49, 121, 122, 125, 178, 179
of trade and investment rules, 17, 18, 47–51, 106, 134
Licensing, 44, 49, 106, 152
Lucent Technologies, 129, 134, 135, 168, 169, 173, 179

Machinery industries, 39, 49, 50, 54, 55
Management practices, 14, 19, 30, 34, 36, 75, 77, 83, 109
Maquiladora industrialization, 26, 45, 63–65, 162, 182
Market access and seeking, 16–18, 25, 39, 74, 75, 84, 102–104, 112, 122, 123, 136–141
Mentoring, environmental, 172, 173
Mergers and acquisitions, 14, 27, 41, 54, 83
Mexico City, 123, 126
Microchips, 89, 90, 93, 94, 123
Microsoft, 83, 152
Microtron, 128, 141
Mining industry, 28–32, 37, 54, 55
Motorola, 69, 117, 128, 135, 163, 173
Multinational corporations
  backward linkages of, 139–148
  benefit limitations of, 102, 182, 188, 189
  environmental practices of, 18, 28–33, 65–69, 96, 97, 164–174
  and FDI, 14–28
  forward linkages of, 150–152
  in late-industrializing countries, 99, 102, 104, 116–119
  Mexico and, 57, 128–131, 178–183
  partnerships with, 23–27, 41, 104, 154, 181
  policy lessons for, 185–191
  proactive utilization of, 38–41, 48, 49, 118, 131, 152, 182, 188

Natural gas industry, 24, 34, 41
NEC, 129, 135
Neoclassical economic theory, 183, 185
Neoliberalism, 47, 49, 51, 102, 141
Non-governmental organizations, 94, 116, 170, 174–176, 179, 191
North American Free Trade Agreement, 43, 48–53, 67, 71, 114, 115, 125, 130–133, 148, 160, 162, 179, 181, 189–191

Occupational health and hazards, 115–117, 160–164, 170–176
Oligopolies, 25, 74, 75, 79, 80, 119
Organization for Economic Cooperation and Development, 28, 29, 44, 52, 57, 65, 115, 168
Outsourcing and offshoring, 36, 37, 60, 61, 72–77, 81–85, 102, 109, 129, 141, 142, 156, 181, 191
Ownership exemption rule, 124, 125

Paris Club, 47, 52
Patent applications, 62, 114
Peripherals, 72, 92, 93, 102–104, 111–114, 122, 123, 129, 130, 136, 144, 145
Personal computers, 72, 73, 81–87, 93, 94, 99, 104, 108, 111–114, 122–131, 141, 150, 151
Peso, valuation of, 46, 47, 52, 55–59, 155, 156, 180
Petroleum industry, 31, 32, 39, 40, 45–47, 74, 132
Philippines, 115, 166
Photolithography, 89, 90
PITEX program, 131, 156, 160, 180
Pollution
  air, 64–67, 94, 96, 115–117, 160, 166–170
  water, 92, 94, 115–117, 160
"Pollution haven" hypothesis, 18, 28, 65, 66, 67, 164–166
Portfolio investment, 14, 27
Prices, 17, 49–51, 58, 78–83, 116
Printers, 49, 72, 136
Procurement, 122, 125, 128, 173
Product development, 17, 19, 27–29, 53, 72–75, 78–80, 109, 110, 147–150
Product differentiation, 78, 81, 119
Production networks
  global, 72–83
  local, 77, 78
  national, 45–48, 160
Property rights, 15, 184
Protectionism, 27, 30, 36, 45

Quotas, 44, 49, 50

Radios, 72, 122
Recycling, 93–97, 161
Regulations, environmental, 30,
    40–42, 67, 68, 88, 94–97, 116–119,
    159–164, 173, 174, 179, 180, 188
Rents, 16, 22, 99
Research and development, 19, 20,
    29, 36, 37, 40, 61, 61, 74, 75, 79,
    80, 84, 103–108, 111, 113, 115,
    118, 119, 122, 125, 131, 135, 140,
    143, 144, 152, 154, 181
Responsible Care program, 35, 170
Risk management, 14, 15, 28, 30, 31,
    77, 81, 83

Safety standards, 88, 96, 97, 162,
    164, 172, 173, 188
Science and technology institutions,
    105–108, 111, 119, 122, 124, 154
SCI-Sanmina, 76, 129, 135, 141, 143,
    161–164, 175
Semiconductors, 90, 93, 94, 105, 124,
    125, 136, 148, 149
Service sector, 16, 17, 76, 77
"Silicon Valley South," 128–152,
    177–183
Skill acquisition, 37–41, 60, 101, 102,
    114, 115, 118, 131, 133, 146, 147,
    151, 178, 179, 189–191
Small and medium-size enterprises,
    52, 113, 154–156
Social peace, 19, 184
Social responsibility, 28, 37, 68, 69,
    96, 163, 191
Software, 72, 108, 147–152, 179
Solectron, 76, 83, 129, 136, 141, 143,
    170, 171, 175
Solvents, 89–91, 170, 171, 179
Specialization, 76–78, 86
Speed to market, 130–133, 137
Spin-off firms, 128–130, 146–150
State-owned enterprises, 45, 51, 108,
    113, 114
Steel industry, 33, 64–67, 88
Stock market, 17, 83, 109, 134, 182

Subcontracting, 60, 106
Subsidiaries, 20, 37, 103
Suppliers, local, 121–126, 131–141,
    140–150, 166, 170–173
Supply chains, 19, 35, 37, 84, 77, 81,
    102–105, 111, 113, 125, 136, 172,
    173, 188
Sustainable development, 13–42,
    48, 49, 53–71, 93–119, 156–159,
    172–191

Tariffs, 16, 17, 44–50, 106, 125, 130,
    131, 134, 156, 157
Tax breaks, 19, 112, 131, 189
Technical support services, 109, 135
Technology centers, 108, 136, 151
Teratogens, 90, 170
Textile industry, 45, 55
Toxicology, 89–91, 170
Trade agreements and policies, 17, 40,
    44–52, 112–117, 156, 157, 179,
    183–191
Trans-national corporations, 57, 58,
    64, 71, 76
Transport industry, 45, 72, 78, 79, 89,
    130–133, 139, 143, 154
Turnkey services, 79, 80

Unisys, 141, 143, 145
United Nations, 100–102, 132, 166
Universities, 106, 108, 113, 144, 148,
    154

Value added, 13, 14, 22, 100, 110,
    122, 123, 132
Value chain, 72–74, 78–81, 100, 109,
    110, 119, 140–150

Wal-Mart, 27, 69
Wang, 123, 128, 141
Washington Consensus, 183, 187
Wastes, 63–65, 88–97, 115–119,
    160–174, 179, 180
World Bank, 32, 49, 165, 173
World Trade Organization, 40, 43,
    44, 51, 53, 56, 114–117, 134, 182,
    187–191